Problems and Solutions
in
Nuclear Physics

Problems and Solutions in Nuclear Physics

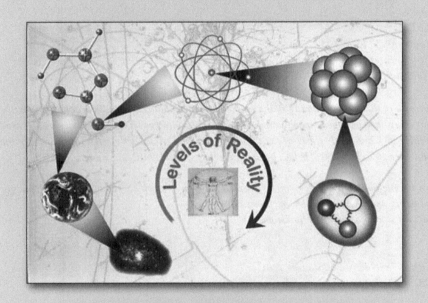

Dr. Mouaiyad M.S. Alabed

iUniverse, Inc.
Bloomington

Problems and Solutions in Nuclear Physics

iUniverse books may be ordered through booksellers or by contacting:

iUniverse
1663 Liberty Drive
Bloomington, IN 47403
www.iuniverse.com
1-800-Authors (1-800-288-4677)

ISBN: 978-1-4759-2606-4 (sc)
ISBN: 978-1-4759-2607-1 (ebk)

Printed in the United States of America

iUniverse rev. date: 05/30/2012

This book is dedicated

To

Imam Ali bin abi Talib
¨Peace be upon him¨

Thanks
To my wife

Nadaa Alsaffar
Without whose inspiring help
this book would never have been completed

PROBLEMS AND SOLUTIONS

Problems

1. From the comparison of the binding energy of the mirrors nuclei (^{11}B) and (^{11}C). Estimate the value (r_0) in the formula of ($R= r_0 A^{1/3}$) for the nuclear radii

2. Define complete (E) and kinetic energy (E_k) of an electron, the given wave length of which equal to (10^{-2}) fm. When the wavelength of the particle which expressed as λ

3. Estimate the diameter of the collision of an (α) particle and nucleus of gold with the bombardment of a target from gold with beam of (α) particles with the kinetic energy of (22Mev). Compare the result with the sum of the nuclear radii of gold and helium

4. Find the density of the nuclear material.

5. What is the velocity and the energy corresponding for a neutron distribution in the thermal equilibrium at ($30°c$)

6. A neutron scattered through an angle of ($30°$) in the centre of mass by collision with a nucleus of ($^{27}Al_{13}$). Find the fractional energy which loss of this particle.

7. A doubly ionized Helium atom is accelerated from rest through a potential difference of (6.10^8) volts in a linear accelerator.

   Calculate:

   a) Its relativistic mass in Kg
   b) Its kinetic energy inMev
   c) Its relativistic velocity.

8. Compare the given wavelengths of an electron and a proton with the identical kinetic energy of (100Mev)

9. Determine the complete energy (E) and the kinetic energy (E_k) of an electron when the wavelength of this electron is equal to (0,01fm)

10. Calculate the wavelength of electrons with (35Gev) as energy

11. Calculate the radius and the approximate mass of the nucleus of ($^{40}Ca_{20}$) and find the approximate density of this nucleus also.

12. To what is equal velocity of particle (v), which's kinetic energy (T) is equal to its rest energy (mc^2)

13. The kinetic energy of a neutron is 4,8Mev collide the nucleus of ($^4He_2$) which will recoil if struck head-on calculate the kinetic energy of ($^4He_2$)

14. An alpha particle has 6,5Mev as a kinetic energy, approach from a foil of gold. Calculate the impact parameter and the cross section in this collision when $\theta = 45^0$, and calculate the distance of the smallest approach in this collision.

15. A particle with 20Mev as a kinetic energy moves in a curvature, find the radius of this curvature when the movement of this particle is perpendicular on a magnetic field of 0,1 W/m^2

16. What is the sum of the kinetic energy of the production of gamma ray, when the energy of this ray is 12Mev?

17. A cyclotron in a research laboratory has a radius of R= 0,5 m. When the magnetic field in this cyclotron is 2,3T.

   a) Find the frequency in this case.
   b) Calculate the kinetic energy of the accelerated protons when they leave this accelerator.

18. What the distance of the intensity of muons beam with a kinetic energy of $E_k = 0.5\,\text{Gev}$ which move in the vacuum, decreases to half of the initial of magnitude?

19. Calculate the wave lengths λ of a proton and of an electron with the kinetic energy E_k of 10Mev

20. A Proton, an electron and a photon have identical (l) wavelength $\lambda = 10^{-9}\,\text{cm}$. What the time t by them necessarily for the flight path of L= 10 m

21. The wavelength of a photon is $\lambda = 3 \cdot 10^{-11}\,\text{cm}$. Calculate the momentum p of this photon

22. Estimate radii of the atomic nuclei ^{27}Al, ^{90}Zr and ^{238}U

23. Estimate the density of a nuclear material when the mass of one nucleon in nucleus is $m_N \approx 1\text{amu} = 1.66 \cdot 10^{-24}$ g

24. Calculate the rest energy for a material of 2,5gram.

25. The nucleus of $^{60}\text{Co}_{27}$ irradiate a photon with the energy of $E_\gamma = 2{,}15.10^{-13}$J How many mass defect is in this isotope?

26. Estimate the nuclear radius of ^{40}Ca with an elastic scattering of electrons when the kinetic energy E_k is 750Mev on the nuclei of ^{40}Ca in the section which observed the diffraction minimum at angle $\theta_{min} = 18°$

27. Calculate the velocity of an electron in the rest state which has been accelerated through a potential difference of 850 kv.

28. A mass of a neutral atom of ^{16}O is 15,9949amu and for ^{15}O is 15,0030amu and for ^{15}N is 15,0001amu What are the energy of the branches of the neutron and of the proton in ^{16}O nucleus

29. An isotope of $^{191}Os_{76}$ has a radioactivity of 2,4 curie uses in a purpose for 8,65days. What is the activity of this isotope at that time

30. What is the minimum energy T_{min} for a deuteron which must have in order as a result of an inelastic scattering on the nucleus of ^{10}B to excite state with the energy E_{ex} of 1.75Mev

31. Compare the mass of a $^{40}Ca_{20}$ nucleus to that of its constituent nucleons.

32. Can we get β^- in the decay process of $^7Be_4$ if the daughter nucleus is $^7Li_3$?

33. A π meson has a mass of 135Mev /C^2. What is the mass in atomic mass units (amu)?

34. a) What is the approximate radius of $^{64}Cu_{29}$ nucleus?
   b) Approximately what is the value of A of a nucleus which has a radius of $3,6.10^{-15}$ m?

35. A neutron with energy of 6Mev collide a $^{239}Pu_{92}$ nucleus. Find the velocity of $^{239}Pu_{92}$ nucleus thrown forward directly.

36. What is the approximate mass of Neptunium $^{239}Np_{93}$ nucleus?

37. How much energy is required to remove a neutron from $^{236}U_{92}$ nucleus?

38. What is the stable nucleus which has a radius equal to the radius of the uranium nucleus?

39. The $^{214}Po_{84}$ nucleus undergoes an alpha decay. What will be the alpha kinetic energy?

40. A proton accelerated in a ring acceleration which has 2T as a magnetic field. Find the periodic frequency when the mass of the proton is $1,67.10^{-27}$ kg.

41. The $^{232}U_{92}$ nucleus irradiates α- particle with the kinetic energy of $E_{k\alpha}$ of 5,32Mev. Find the daughter nucleus, which the value of the approximate atomic mass (with amu units) for the daughter atom.

42. When $^{23}Ne_{10}$ (mass= 22,9945amu) decay to $^{23}Na_{11}$ (mass = 22,9898amu), what is the maximum kinetic energy of the emitted electron? What is the minimum energy and the energy of the neutrino in each case?

43. Proof how we can express the atomic mass unit for

44. Compare between the mass of helium ($^4He_2$) nucleus and a nucleus which has some nucleons.

45. A mass with 10grams. Calculate its energy with ev unit, which fall from the origin under effect of gravitational force, with the distance of 15cm. Compare this result with fall hydrogen molecule energy through the same distance.

46. What is the approximate mass of Neptunium $^{239}Np_{93}$?

47. a) What is the approximate radius of $^{64}Cu_{29}$ nucleus?
    b) Approximately what is the value of A for a nucleus whose radius is $3,6.10^{-15}m$?

48. How long will it take a 1Mev neutron to cross a $^{238}U_{92}$ nucleus?

49. The velocity of alpha particles from plutonium $^{239}Pu_{94}$ is $1,5.10^7$ m/sec. What is the velocity of the recoil atom?

50. What is the fractional increase in mass of a 10Mev proton?

51. What is the neutron temperature corresponds to 8ev?

52. Estimate the value of the charge number Z, with which the nuclei become unstable with the relation to the spontaneous decay

53. Estimate a change in the binding energy of a heavy nucleus in the fission of this nucleus into two identical nuclei of splinter. Consider the case A= 240, Z= of 92

54. a) Calculate the total binding energy and the binding energy per nucleon of $^3H_1$,$^4He_2$,and $^{31}Si_{14}$.
    b) Find the energy would be involved by removing the nucleons in Helium to infinity?

55. Determine kinetic energies $E_{k\alpha}$ of α - particles, which are formed with α-disintegration $^{212}Bi_{83}$ to the excited states of nucleus $^{208}Tl_{81}$ with energies 0.49Mev and 0,61Mev. The binding energy E_{BE} of $^{212}Bi_{83}$ nucleus is 1654,32Mev, of $^{208}Tl_{81}$ nucleus is 1632,23Mev and of α - particle is 28,30Mev.

56. Determine the recoil energy of nucleus $^7Li_3$, which is formed with e capture in the nucleus of $^7Be_4$

57. Determine the kinetic energy of residual nucleus with β⁻ nuclear decomposition $^{64}Cu_{29}$ ($^{64}Cu_{29}$ → $^{64}Zn_{30}$+ e+ $\acute{v}_e$) when

    1) energy of antineutrino $T_{\dot{v}}$= 0,
    2) energy of electron T_e= 0.

    The nuclear binding energy of $^{64}Cu_{29}$ is 559,32Mev and for $^{64}Zn_{30}$ is 559,12Mev.

58. Find the separation energy of an alpha particle from $^{12}C_6$

59. The masses of a neutron and of a proton in the energy units are respectively m_n= 939,6Mev and m_p= 938,3Mev. Determine nuclear mass $^2H_1$ in the energy units, if the binding energy of deuteron is E_{BE}= 2,2Mev

60. Determine the specific nuclear binding energy of $^{16}O_8$ nucleus When the mass of the neutral atom of $^{16}O_8$ M is 15,9949amu

61. Calculate the energies of the neutrons branches in even-even isotopes $^{38}Ca_{20}$, $^{40}Ca_{20}$, $^{48}Ca_{20}$ with the aid of Weizsacker's formula

62. Calculate radii of the mirror nuclei $^{23}Na_{11}$, $^{23}Mg_{12}$. When the binding energy of these nuclei are: $E_{BE}(^{23}Na_{11})= 186,56$ Mev, $E_{BE}(^{23}Mg_{12})= 181,72$ Mev

63. The ^{27}Si nucleus as a result of β^+- disintegration
$$^{27}Si_{14} \quad \rightarrow \quad ^{27}Al_{13} + e^+ + \nu_e$$
Passes into the ^{27}Al as one of mirror nuclei. Maximum energy of positrons is 3.48Mev. Estimate a radius of these nuclei

64. By using the Weizsacker's formula, obtain the relationship for enumerating the energy of spontaneous fission for two identical splinters and design energy of symmetrical nuclear fission of $^{238}U_{92}$

65. Does appear reaction $^6Li_3(d,\alpha)^4He_2$ of endothermic or exothermic, when we are given the specific nuclear binding energy inMev: $\varepsilon(d)= 1,11$; $\varepsilon(\alpha)= 7,08$; $\varepsilon(^6Li_3)= 5,33$

66. What are the total binding energy and the average of the binding energy per particle of $^{235}U_{92}$ nucleus?

67. What is the binding energy of the isotope $^{210}Pb_{82}$? What is the binding energy per nucleon?

68. Calculate the binding energy of the last neutron in a $^{12}C_6$ nucleus. (Hint: compare the mass of $^{12}C_6$ with that of $^{11}C_6+ {}^1n_0$ use appendix).

69. Calculate the total binding energy and the binding energy per nucleon for $^6Li_3$ Use appendix.

70. What is the binding energy of the isotope $^{210}Pb_{82}$? What is the binding energy per nucleon for this isotope?

71. What are the total binding energy and average binding energy per particle of nucleus $^{235}U_{92}$?

72. A nitrogen $^{14}N_7$ nucleus absorbs a deuterium $^2H_1$ nucleus during a nuclear reaction. What is the name, atomic number and nucleon number of the compound nucleus and find the type of this reaction.

73. Two isotopes of the same element have the same binding energy. One isotope contains two more neutrons than the other. What is the difference between the atomic masses (in atomic mass units) of these isotopes?

74. Calculate the binding energy of $^{40}Ca_{20}$ nuclide with using the following scheme.

75. Calculate the: a) Total binding energy of $^{238}U_{92}$ nucleus.

    b) Total binding energy of $^{107}Ag_{47}$ nucleus.

76. a) Determine that the nucleus $^8Be_4$ is instable to decay to two alpha particles,
    b) Is the isotope $^{12}C_6$ can decay to three alpha particles? If is it yes, why? And if is it no, why?
77. Is the following reaction occurs or not? Why?
    $$^{11}C_6 \quad \rightarrow \quad ^{10}B_5 + p$$

78. Find the total binding energy and the average binding energy per nucleon for the $^{90}Sr_{38}$ nucleus.

79. Find the binding energy of the isotope $^{152}Sm_{62}$ and find the binding energy of this isotope per nucleon?

80. Calculate the binding energy with the nucleon units to the nucleus of $^{12}C_6$.

81. Calculate the binding energy of the last neutron in the nucleus of $^{12}C_6$ (compare the mass of $^{12}C_6$ with the produced mass to $^{11}C_6 + {}^1n_0$).

82. The isotope of $^{22}Na_{11}$ in a tissue an approximate 0,01grams per kilogram tissue ,and the isotope of $^{40}K_{19}$ in this tissue at an approximate 0,008grams per kilogram total tissue. Calculate the radioactivity of two isotopes in this tissue, when the mass of this tissue is 2 kg.

83. What is the mass of 1 µCi of $^{98}Tc_{43}$ ($T_{1/2} = 4,2.10^6yr$)?

84. A 0,018 µCi of a sample of $^{32}P_{15}$ is injected into an animal for tracer studies. If a Geiger counter intercepts 20 percent of the emitted β particles and is 90 percent efficient in counting them, what will be the counting rate?

85. A sample of the nucleus of $^{14}C_6$ which has 1.10^{21} nuclei and its half-life is 5730 yr. Calculate its radioactivity.

86. Obtain the relationship between the half-life period, disintegration probability and the average of half-life

87. Find the kinetic energy of α- particle and recoil nuclei in the disintegration of radium

88. Determine the activity of the preparation of gold – 198($^{198}Au_{79}$), induced during irradiation of the model of gold - 197($^{197}Au_{79}$) with mass of 0.1g in the thermal neutron flux $10^{12}cm^{-2} s^{-1}$ for 1 hour. The cross section of auric activation by thermal neutrons is 97 barns

89. They are given the difference of the masses of atoms

Δ ($^{114}Cd_{48}$)= -90.021Mev, $\Delta(^{114}In_{49})$= -88.379Mev and $\Delta(^{114}Sn_{50})$= -90.558Mev.
Determine the possible forms of β- nuclear decomposition of ($^{114}In_{49}$)

90. For $^{17}Ne_{10}$ nucleus, determine the maximum energy of the being late protons, which depart from the $^{17}F_9$ nucleus, which is formed as a result of e-capture on nucleus $^{17}Ne_{10}$

91. Find the separation energy of a neutron and of a proton from the nucleus of $^{12}C_6$

92. How many times is the number of nuclear decompositions of radioactive iodine $^{131}I_{53}$ during the first day more than the number of disintegrations during the second day? The period of the half-life of $^{131}I_{53}$ isotope equal to193 hours.

93. Determine the isolated energy W, by 1 mg of preparation of $^{210}Po_{84}$ in the time which equal to the average of the life, if the energy E is 5.4Mev separated with one report of the disintegration.

94. Determine upper boundary of the age of the Earth, considering that entire existing on the earth, $^{40}Ar_{18}$ was formed from $^{40}K_{19}$ as a result of e-capture. At present to every 300 atoms of $^{40}Ar_{18}$ is fallen one atom $^{40}K_{19}$

95. As a result of α- disintegration radium $^{226}Ra_{88}$ is converted into radon $^{222}Rn_{86}$. What is the volume of radon with the standard conditions which located in the equilibrium with 1 g of radium? The period of half-life $^{226}Ra_{88}$; $T_{1/2}$ is 1600 years, and for $^{222}Rn_{86}$ is $T_{1/2}= 3.82$ days

96. A proton collide a target of the natural Lithium. When the activity was 140Bq, Which the radioactive source of this activity, when the time was 106 days to reduce the activity to 35Bq?

97. A plate from ^{55}Mn with a thickness d= 0.2cm, which radiated the beam of neutrons J in the time of $t_{act}= 12$ min, if $t_{oxl}= 100$ min and the activity was 2800 Bq, Calculate the intensity of neutron beam J, when the cross section was σ= 0.48 barns, and the density of the plate is ρ= 7.43 g/cm³.

98. How much energy will be released when two deuterons. Combine to form an alpha-particle?

99. The isotope of $^{218}Po_{84}$ can decay by either α or β⁻ emission. What is the energy release in each case? The mass of $^{218}Po_{84}$ is 218,008965amu

100. The nuclide of $^{32}P_{15}$ decay by emitting an electron whose maximum kinetic energy can be 1,71Mev.

    (a) What is the daughter nucleus?
    (b) What is its atomic mass in (amu)?

101. Show that the decay $^{11}C_6 \rightarrow {}^{10}B_5 + {}^{1}P_1$ is not possible because the energy would not be conserved.

102. (a) Show that the nucleus of $^{8}Be_4$ (m= 8,005308amu) is unstable to decay into two α particles.
    (b) Is $^{12}C_6$ stable against decay into three α particles? Show why or why not.

103. The maximum permissible concentration of radium emanation in the air for continuous exposure is 10^{-8} micro curies per milliliter of the air. For this concentration what is the radon content of 1 ml of standard air in percent by the mass and in percent by the volume? The density of the air at standard conditions is $1,2929. 10^{-3}$ gm /cm³

104. What is the weight in grams of 1 curie of 5,3year of $^{137}Cs_{55}$?

105. What is the activity of a gold foil that has been irradiated for a period of 10 hours, assuming a constant rate of the formation of the radioisotope in the reactor of 10^9 atoms per sec, if its half-half is 53,3days.

106. How much energy is required to remove a neutron from $^{236}U_{92}$ nucleus?

107. A sample has a $^{14}C_6$. The activity of 0,0061Bq per gram of carbon. Find the age of the sample, assuming that the activity per gram of carbon in a living organism has been constant at a value of 0,23Bq.

108. The practical limit to ages that can be determined by radiocarbon dating is about 41000yr. In a 41000 yr-old sample, what percentage of the original $^{14}C_6$ atoms remains?

109. Bones of the woolly mammoth have been found in North America. The youngest of these bones has a $^{14}C_6$ activity per gram of carbon that is about 21% of what was present in the live animal. How long ago (in years) did this animal disappear from North America?

110. What is the mass (in grams) of krypton $^{92}Kr_{36}$($T_{1/2}$= 1,84 s) which has the same activity as 2gram of Xenon $^{140}Xe_{54}$ ($T_{1/2}$=13,6 s)?

111. To see why one curie of activity was chosen to be $3,7.10^{10}$ Bq. Determine the activity of one gram of radium $^{226}Ra_{88}$ ($T_{1/2}$= $1,6.10^3$ yr).

112. What is the rest energy with Mev/c^2 of an α-particle.

113. Calculate the decay energy of uranium nucleus ($^{232}U_{92}$) which decays to thorium $^{228}Th_{90}$ and alpha particle.

114. The 2 microgram of Radium nucleus ($^A Ra_{88}$) which has the atomic number of 88. Its nucleus irradiates an alpha particle and follows gamma rays. The half-life of radium is 1600 years. Calculate the number of alpha particles in the decay process

115. What is the mass and the volume under standard conditions of two curie of radium emanation (radon)?

116. What the ratio at the time of the earth formation when the ratio of the number of $^{238}U_{92}$(the half-life is $4,5.10^9$ years) and

the number of $^{235}U_{92}$(the half-life is $7,1.10^8$ years) when this ratio between the two isotopes is 140($^{238}U_{92}$ / $^{235}U_{92}$ = 140) and the earth is 4.10^9 years old.

117. A sample of phosphorus contains 8% atoms of phosphorus($^{32}P_{15}$) with the radioactivity of 0,023 micro curie.

   a) Find the number of atoms of this isotope in this sample.
   b) Find the total mass of this sample ($T_{1/2}$= 14,262days).

118. How many alpha particles are emitted per second by 3 microgram of radium?

119. How many disintegrations occur each second in 0,5gram of $^{243}Am_{95}$, and what is the number of curies in this isotope ?

120. What is the energy of an alpha particle which released in the following disintegration?

$^{210}Po_{84}$ → $^{206}Pb_{82}$+ α+ Q(?)

121. In 9 days the number of radioactive nuclei decreases to one-eight the number present initially. What is the half-life (in days) of the material?

122. The number of the radioactive nuclei present at the start of an experiment is $4,6.10^{15}$. The number present twenty days later is $8,14.10^{14}$. What is the half-life (in days) of the nuclei?

123. A radioactive material decay of 1280 dis/minute. After 6 hours its decay becomes 320dis/min. What is the half-life for this material?

124. What is the decay constant for Uranium ($^{238}U_{92}$) which has a half-life of $4,468.10^9$ year?

125. What is the activity for $^{14}C_6$ nucleus which has $5,6.10^{20}$ nuclei?

126. The isotope $^{243}Am_{95}$ has a half-life of 7380 y.

   a) How many nucleus in this isotope, when its activity is $5,5.10^{15}$ decay/sec.

   b) What is the activity of this isotope after 500y and

   c) What is the activity after 7000y?

127. (2)μg of $^{235}U_{92}$,which has a half-life of $7,038.10^8$y.

   a) How many nuclei in this nucleus?

   b) What is the initial activity for this isotope?

   c) What is the activity after 5 million years?

   d) Which the time must be the activity 1/10 of its initial activity?

128. The boney form has 9,2grams of $^{14}C_6$. Its radioactivity is 1,6Bq. How many years old this boney form?

129. How many nucleus of $^{238}U_{92}$ in the stone, when the activity of this isotope is of $1,8.10^4$ decay / sec?

130. The nuclide $^{124}Cs_{55}$ has half-life of 30,8 sec.

   a) If its 9,5 microgram in the initial state , how many nucleons in this state?

   b) How many nucleons will be after 2 minutes?

   c) What is the activity at this time?

   d) After which the time decreases the activity to the less than one in the second?

131. Calculate the activity for the pure mass of 3,5 microgram of the nuclide $^{32}P_{15}$, when the half-life of this nuclide is $1,23.10^6$ sec.

132. The radioactive sample of $^{210}Po_{84}$ isotope is 0,4milligram. When the half-life of this isotope is 138,376 days. Which mass will be stay from this material after one year?

133. How many alpha particles and beta particles which emitted in the series of decay from $^{238}U_{92}$ to $^{206}Pb_{82}$?

134. How much energy is required to remove a neutron from $^{24}Na_{11}$ nucleus?

135. If the activity of a radioactive substance is initially 398dis /min and two days later it is 285 dis / min, what is the activity four days later still, or six days after the start? Give your answer in dis / min.

136. To make the dial of a watch glow in the dark,1.10^{-9}kg of radium $^{226}Ra_{88}$ is used. The half-life of this isotope is $1,6.10^{3}$years. How many kilograms of radium disappear while the watch is in use for fifty years?

137. Calculate the width of the first excited energy level of an isotope, when the energy of this level is 0,87Mev above ground level, and the half-life of this isotope is $1,73.10^{-4}$ μsec.

138. How many disintegrations occur each second in 0,5g of a sample of $^{243}Am_{95}$? What is the number of curies in this mass?

139. Two milligram of Actinium-227($^{227}Ac_{89}$) half-life is 21,773 years, is allowed to decay for 1 year, what is the activity at the end of that time?

140. How many nucleus of $^{249}Cf_{98}$ isotope that will accumulate in 170years with a generation rate of 10^{14} nucleus/sec, and how many grams does this comprise?

141. How many disintegrations occur each second in a 0,5g of a sample of $^{245}Cm_{96}$?

142. An isotope of ($^{191}Os_{76}$) has a radioactivity of 2,4curie uses in a purpose for 8,65 days. What is the activity at that time?

143. The isotope of $^{35}P_{15}$ has a half-life of 14,262 days.

   a) How long the time must take $0.5N_0$ (N_0:the original number of nucleus) to decay?
   b) Find the mean life (τ) of this isotope.

144. What is the fraction of radium emission as a sample of radon will disintegrate in two days? $^{222}Rn_{86}$ ($T_{1/2}$=3,82 days).

145. What is the fraction of the original amount remains a radioactive element disintegrates for a period of time equal to its average life?

146. How much $^{238}U_{92}$ which contains after disintegrating for a period equal to its average life, when a piece of this element is 4 gm.

147. What is the age of the uranium ore if now contains 0,5gm of $^{206}Pb_{82}$ for each gram of $^{238}U_{92}$?

148. What is the mass of one curie of?

   a) $^{235}U_{92}$
   b) $^{210}Po_{84}$
   c) $^{239}Pu_{94}$

   T_U= 7,038.10^{8}years
   T_{Po}= 138,376 days
   T_{pu}= 24,119 years.

149. a) Assuming that there are 10 tons of $^{238}U_{92}$ and 15 tons of $^{232}Th_{90}$ uniformly distributed in the first meter of depth under each square kilometre of the surface of the earth. Find the combined activity of these two radioelements in micro curies in the cubic meter of soil under each square meter of surface.
   b) Assuming radioactive equilibrium, find the mass of $^{226}Ra_{88}$ in one cubic foot of soil.

150. What is the number of alpha decays in one gram sample of Californium $^{249}Cf_{98}$ ($T_{1/2}$= 351years) in 50 years, and in 300 years.

151. Calculate the activity of 5gram of $^{249}Cf_{98}$ and the specific activity of this isotope.

152. The isotope of $^{60}Co_{27}$ responsible for the activity of beta $2,63.10^8$dis/sec, which makes up 0,014% of natural mixture. When the mass occurring cobalt is 45mg naturally. What the half-life of this isotope.

153. Iodine $^{131}I_{53}$ used in diagnostic and therapeutic techniques in the treatment of thyroid disorders. This isotope has a half-life of 8,04days. What percentage of an initial sample of this isotope remains after 30 days?

154. Find the separation energy of 2α particles from Radium $^{226}Ra_{88}$?

155. How many Mev for the separation energy of 1n from $^{24}Na_{11}$?

156. Estimate the nuclear binding energy of $^{12}C_6$ according to Weizsacker formula and to compare result with the same value, obtained from the experimental data about the masses.

157. A radioactive material produces 1280 decay per minute at one time, and 6h later produces 320 decays per minute. What is the half-life of this material?

158. What is the fraction of a sample of $^{68}Ge_{32}$, whose half-life is about 9 months, will remain after 4,5yr?

159. $^{124}Cs_{55}$ has a half-life of 30,8 sec.

   a) If we have 9,5 µg initially, how many nuclei are present?
   b) How many are present 2 min later?
   c) What is the activity at this time?
   d) After how much time will the activity drops to less than about 1 per second?

160. The activity of a sample of $^{35}S_{16}$ ($T_{1/2}= 7,56.10^6$ sec) is $6,8.10^6$ decays per second. What is the mass of the sample present?

161. A radioactive nuclide produces 2880 decays per minute at one time, and 1,6h later produces 820 decays per minute. What is the half-life of the nuclide?

162. Proton with the kinetic energy E_k of 2Mev strikes stationary nucleus $^{197}Au_{79}$. Determine the differential scattering cross-section $d\sigma/d\Omega$ at an angle $\theta= 60°$ As the value of the differential scattering cross-section will change, if we as the scattering nucleus select $^{27}Al_{13}$

163. Calculate the scattering cross section of α - particles with the kinetic energy

    $E_k = 5Mev$ by the coulomb field of $^{208}Pb_{82}$ nucleus at angles are more than 90^0

164. Calculate the differential cross section $d\sigma/d\Omega$ of an elastic proton scatterings on the nuclei of gold $^{197}Au_{79}$ at the angle of 15°, if it is known that for the section of the irradiation of target with a thickness d= 7 mg/cm^2 by protons with the summary charge Q=1 nCuloumb to the detector with an area S= 0.5cm^2, located at a distance of l= 30cm from the target, it fell that $\Delta N= 1,97 \cdot 10^5$ of the elastic scattered protons. Compare the experimentally measured section from Rutherford differential cross-section.

165. Determine the cross-section of the reaction of $^{31}P_{15}(n,p)^{31}Si_{14}$, if it is known that after the irradiation target $^{31}P_{15}$ with a thickness of d= 1 g/cm^2 in the neutron flux J= $2\cdot10^{10}$ c^{-1}·cm^{-2} in period $t_{ir}= 4$ h, its β- activity I, measured after the time of $t_{ir}= 1$ hour after the end of irradiation, proved to be $I(t_{pr})= 3.9\cdot10^6$ decay/s. Period of the half-life $T_{1/2}(^{31}Si_{14})= 157.3$ min

166. Calculate the thermal fission cross section for an enriched uranium mixture which contains 2% of $^{235}U_{92}$ atoms.

167. Calculate the fission rate for $^{235}U_{92}$ required to produce 2 watt and the amount of the energy that is released in the complete fission of 1800grams of this isotope.

168. The isotope of $^{235}U_{92}$ has a half-life of $7,038.10^8$ years for fission. Estimate the rate of the fission for 1gram of $^{235}U_{92}$.

169. The cross section for the (n,α) reaction with boron follows the 1/v law. If the cross section for 40 eV neutrons is 16 barns, calculate the cross section for neutrons with 0,024eV.

170. Calculate the value of the integral section when the differential cross section of the reaction is $d\sigma/d\Omega$ at the angle of 90^0 composes 10mb/sr

171. Calculate the scattering cross section of an α - particle with the energy of 3Mev in the coulomb field of $^{238}U_{92}$ nucleus in the angular interval from 150^0 to 170^0

172. The nucleus of $^{10}B_5$ from the excited state with the energy of 0.72Mev decomposed by the emission of γ- quantum with the period of half-life

    $T_{1/2} = 6.7 \cdot 10^{-10}$s. Estimate uncertainty energy ΔE of that emitted γ- quantum

173. Gold plate with a thickness of l= 1 micrometer irradiates by a beam of α- particles with the density of flow j= 10^5 particle/ cm^2.s; Kinetic energy of α- particles E_k is 5Mev. How much α- particles per unit of solid angle does fall per second to the detector, which located at the angle θ= 170° to the axis of bundle? Area of the bundle spot on target S is 1cm^2

174. Determine the upper boundary of the positrons spectrum, emitted when β^+ - nuclear decomposition of $^{27}Si_{14}$. The energy of β^+disintegration, by using values of the masses of atoms

175. **Transfer several nuclear reactions, in which can be formed isotope $^8Be_4$**

176. **What is the minimum kinetic energy in the laboratory system $E_{k\,min}$ which must have a neutron so that would become possible reaction $^{16}O_8(n,\alpha)^{13}C_6$**

177. **Determine the thresholds T_{th} of the reactions of the $^{12}C_6$ photo disintegration:**

    1. $\gamma + {}^{12}C_6 \rightarrow {}^{11}C_6 + n$
    2. $\gamma + {}^{12}C_6 \rightarrow {}^{11}B_5 + p$
    3. $\gamma + {}^{14}C_6 \rightarrow {}^{12}C_6 + n + n$

178. **Determine the thresholds of the two reactions: $^7Li_3(p,\alpha)^4He_2$ and $^7Li_3(p,\gamma)^8Be_4$**

179. **What is the minimum energy which a proton must have; that would become possible the following reaction?**
    $p + d \rightarrow p + p + n$

180. **Is the reactions are possible**

    $\alpha + {}^7Li_3 \rightarrow {}^{10}B_5 + n$
    $\alpha + {}^{12}C_6 \rightarrow {}^{14}N_7 + d$

181. **Calculate the energy Q of the following reactions and, identify the particle of X**

182. **Calculate the threshold of the following reaction**

    $^{14}N_7 + \alpha \rightarrow {}^{17}O_8 + P$ in two cases, if the collide particle is:
    1) An α- particle
    2) Nucleus $^{14}N_7$, and the energy of the reaction is Q= 1.15Mev

183. Find the types of the energies and thresholds of the following reaction

$^{32}S_{16}(\gamma,\alpha)^{28}Si_{14}$
$^{7}Li_{3}(p,n)^{7}Be_{4}$
$^{4}He_{2}(\alpha,p)^{7}Li_{3}$

184. What is the energy of γ- quant when a slow neutron collide the $^{7}Li_{3}$ nucleus?

185. Find the kinetic energy of $^{9}Be_{4}$ nucleus in the laboratory system which is formed with the threshold value of the neutron energy in the reaction of $^{12}C(n,\alpha)^{9}Be_{4}$

186. The sign of this result is negative, that means the direction of the nucleus is inverse of alpha particle
Which the radioactive isotope were formed in these nuclear reactions

$^{11}B(t,2p)$?, $^{11}B(t,2n)$?, $^{11}B(\alpha,t)$

187. Alpha particle with the kinetic energy E_k of 20Mev experiences elastic head-on collision with the nucleus of $^{12}C_6$. Determine the kinetic energy in hp of nucleus $^{12}C_6$ E_{kC} after collision

188. Determine the maximum and the minimum energy of nuclei of $^{7}Be_4$, that is formed in the reaction of $^{14}N(n,p)^{14}C$ when Q= 1,65Mev under the action of the accelerated protons with the energy of E_{kn}= 6Mev.

189. The energy of alpha particles is $E_{k\alpha}$= 10Mev, which interact with stationary nucleus of $^{7}Li_3$. Determine the values of pulses in C.M. system when the form of the reaction is $^{7}Li(\alpha,n)^{10}B$

190. Determine the energy of the protons which observed in the reaction of $^{32}S(\alpha,p)^{35}Cl$ at angles of 0^0 and 90^0 with E= 8Mev. The excited states $^{35}Cl_{17}$ (1,219; 1,763; 2,646; 2,694; 3,003; 3,163)Mev

The situation will be excited on the beam of α- particles with energy of 5Mev

191. The kinetic energy of a proton is E_{kp} = 7Mev collide the nucleus of $^1H_1$ and the proton scattered with an elastic scattering which the kinetic energy E_k of the B nucleus and scattering angle θ_B of the recoil nucleus $^1H_1$ if the scattering angle of $^1H_1$ is θ_b = 40°

192. Determine the energy of a neutron E_{kn} which scattered at the angle of 90° in the neutron generator and uses the deuterons which accelerated to the energy of E_{kd} = 0,5Mev, when the reaction is t(d,n)α.

193. How much the energy of two deuterons which will be released in a reaction. Combine to form an alpha- particle?

194. Calculate the power which will result from a fission rate of 3. 10^{12} per second.

195. Calculate ζ for $^9Be_4$ – moderator.

196. How much $^{235}U_{92}$ has been consumed in a reactor which has operated for 8 years at an average power of 180 watts?

197. How many fissions per second are required to produce one watt of power?

198. How much power is generated by the fission of one gram of $^{235}U_{92}$ per 8 hour.

199. Calculate the number of the collisions required to reduce the fast fission neutrons with an average initial energy of 2Mev to an energy

E_t = 0,024 ev in Beryllium (as a moderated assembly for Be_4 is ζ = 0,209.

200. What would be the energy of the neutrons that have 45 collisions with Be_4 nuclei which start with an initial energy of 2Mev?

201. What is the maximum fractional energy loss for the neutrons in the collisions with $^{39}K_{19}$ nuclei?

202. Calculate the average number of the fission neutrons per neutron absorbed in a uranium mixture which contains $^{235}U_{92}$ and $^{238}U_{92}$ isotopes in 1:12 ratio.

203. What is the ratio of the released energy from the nuclear fission of $^{235}U_{92}$ to its rest energy, when the released energy in this process is about 200Mev.

204. What is the combined mass of the two fragments and there energy in the nuclear fission process to the $^{235}U_{92}$ nucleus when the released energy during this process is about 225Mev.

205. The nucleus $^{235}U_{92}$ absorbs a thermal neutron and make a nuclear fission process into Rubidium $^{93}Rb_{37}$ and Cesium $^{141}Cs_{55}$. What are the nucleons produces by this process and how many are there?

206. Determine the number of the neutrons which released during the following fission reaction:

$$^{235}U_{92} + {}^1n_0 \rightarrow {}^{133}Sb_{51} + {}^{99}Nb_{41} + ?({}^1n_0)$$

207. Consider the induced nuclear reaction:

$$^2H_1 + {}^{14}N_7 \rightarrow {}^{12}C_6 + {}^4He_2$$

And determine the energy in Mev which released when $^{12}C_6$ and $^4He_2$ nuclei are formed in this manner.

208. During a nuclear reaction, an unknown particle is absorbed by a copper $^{63}Cu_{29}$ nucleus, and the reaction products are $^{62}Cu_{29}$, a

neutron and a proton. What is the name, the atomic number and the nucleon number of the compound nucleus?

209. Write the reactions below in the shorthand form:

a) $^1n_0 + ^{14}N_7 \rightarrow ^{14}C_6 + ^1H_1$
b) $^1n_0 + ^{238}U_{92} \rightarrow ^{239}U_{92} + \gamma$
c) $^1n_0 + ^{24}Mg_{12} \rightarrow ^{23}Na_{11} + ^2H_1$

210. Complete the following nuclear reactions, assuming that the unknown quantity signified by the question mark is a single entity:

a) $^{43}Ca_{24}\,(\alpha\,,?\,)\,^{46}Sc_{21}$
b) $^9Be_4(?\,,n\,)\,^{12}C_6$
c) $^9Be_4\,(p\,,\alpha\,)\,?$
d) $?\,(\alpha\,,p\,)\,^{17}O_8$
e) $^{55}Mn_{25}(\,n\,,\gamma\,)\,?$

211. What is the nucleon number A in the reaction of:

$^{27}Al_{13}\,(\alpha\,,n\,)\,^AP_{15}\,?$

212. What is the atomic number Z and the element X in the reaction $^{10}B_5(\alpha,p)\,^{13}X_Z\,?$

213. In the following reaction:

$^1n_0 + ^{10}B_5 \rightarrow ^7Li_3 + ^4He_2$

Produce $^7Li_3$ nucleus with an alpha particle. The kinetic energy of $^1n_0$ is slow, and the $^{10}B_5$ nucleus in the ground state. The kinetic energy of alpha particle is $9,3.10^6$ m/s. Calculate

a) The kinetic energy of $^7Li_3$ nucleus and
b) The produced energy of this reaction.

214. In the nuclear medicine, the maximum permissible concentration of the radium emission for the continuous exposure in the environment is

10^{-7} micro curies per milliliter of the air. What is the radon content of milliliter for this concentration of the standard air?

215. Is the energy enough to occur this reaction?

$$^{13}C_6 + {}^1H_1 \rightarrow {}^{13}N_7 + {}^1n_0$$

216. Calculate the energy which released in the following reaction.

$$^{14}C_7 \rightarrow {}^{14}N_7 + \beta^-$$

217. The nucleus of $^{32}P_{15}$ decay according to the following reaction, what is the daughter nucleus and its atomic mass with amu units, when the kinetic energy of the electron is maximum and it is of 1,71Mev?

$$^{32}P_{15} \rightarrow X_{16} + \beta^-$$

218. What is the released energy in the following reaction, when the nucleus decays with an alpha particle or with a beta particle? The mass of $^{218}Po_{84}$ nucleus is 218,008965amu

$$^{218}Po_{84} \rightarrow {}^{214}X_{82} + {}^4He_2(\alpha) + Q$$

$$^{218}Po_{84} \rightarrow {}^{218}X_{85} + \beta^- + Q$$

219. Calculate the released energy in an electron capture process by Beryllium nucleus.

220. How many collisions with a carbon nucleus are required to reduce the energy of a fast neutron from 2Mev to 0,04eV?

221. How many collisions must make before neutrons speed reduced below the average speed of a hydrogen molecule at $20^{\circ}C$ when these neutrons moves with an initial speed of 10^{9}cm/sec through the hydrogen gas and one of them loses exactly 1/4 of its speed in the each collision.

222. Complete the following reactions:

a) $^{211}Pb_{82}$ $\rightarrow$ $^{211}Bi_{83}$ + ?
b) $^{11}C_{6}$ $\rightarrow$ $^{11}B_{5}$ + ?
c) $^{231}Th_{90}{}^{*}$ $\rightarrow$ $^{231}Th_{90}$ + ?
d) $^{210}Po_{84}$ $\rightarrow$ $^{206}Pb_{82}$ + ?
e) $^{2}D_{1} + {}^{2}D_{1}$ $\rightarrow$? + $^{1}n_{0}$ + Q
f) $^{9}Be_{4} + {}^{4}He_{2}$ $\rightarrow$? + $^{1}n_{0}$
g) $^{31}P_{15} + {}^{1}n_{0}$ $\rightarrow$ $^{32}P_{15}$ $\rightarrow$? + $^{0}e_{-1}$
h) $^{24}Na_{11}$ $\rightarrow$ $^{24}Na_{12}$ + ?

223. In the nuclear fusion operation two nucleus of a deuterium make the following reaction:

$$^{2}H_{1} + {}^{2}H_{1} \rightarrow {}^{3}He_{2} + {}^{1}n_{0}$$

Calculate the value of the energy which release in this operation.

224. Two deuterons ($^{2}H_{1}$ nucleus) combine to form an alpha-particle. Which the energy will be released in this reaction?

225. Find the Q-value of the following nuclear reaction:

$$^{231}Pa_{91} \rightarrow {}^{227}Ac_{89} + {}^{4}He_{2} + Q$$

When the energy of an alpha particle is 6,2Mev.

226. What is the energy of gamma rays which accompanies the decay of $^{40}K_{19}$ as the following:

$$^{40}K_{19} \rightarrow {}^{0}e_{-1} + {}^{40}Ca_{20}$$

227. Calculate the neutron velocity that corresponds to the energy of 4Mev in the fission process of $^{239}Pu_{94}$ nucleus.

228. A device used in a radiation therapy for cancer contains 0,5 g of cobalt $^{60}Co_{27}$. The half-life of this isotope is 5,27years. Determine the activity of the radioactive material.

229. Find the Q- value of the nuclear reactions as:

$^{226}Ra_{88} \rightarrow {}^{222}Rn_{86} + {}^{4}He_{2} + Q$, when the energy of an alpha particle is 6,2Mev.

230. In the fission of $^{235}U_{92}$ nucleus with two fragments are 4 neutrons and 2 beta particles. Find the fragments when these are equal.

231. Show that the kinetic energy of a neutron in the nuclear reaction of alpha particle and Beryllium is 5,7Mev, When the product nucleus in this reaction is $^{12}C_{6}$.

232. A neutron with energy of 4Mev forward to the graphite and collide with it. How many collisions are required for this particle to loose on the average of 95% of an initial energy?

233. How many collisions with a carbon nucleus are required to reduce the energy of a fast neutron from 2Mev to 1/30ev?

234. What the angle of θ_{n} relative to the direction of a proton beam will collide neutrons with this energy. When the neutron production are used the reaction of $^{7}Li_{3}(p,n)^{7}Be_{4}$, and the energy of the protons is $E_{kp} = 5$Mev, and the neutrons with the energy E_{kn} of 1.75Mev

235. What the initial mass of $^{235}U_{92}$ is required to operate a 500-MW reactor for one year?

236. Calculate the energy which released per gram of a fuel for the reaction:

$$^2H_1 + {}^2H_1 \rightarrow {}^3H_1 + {}^1H_1 \text{, When Q= 4,03Mev}$$

237. Show that the energy released when the two deuterium nuclei fuse to form $^3He_2$ with the release of a neutron is 3,27Mev.

238. Quant of gamma ray transfer through the mass of $6.\ 10^{-3}$ kg of the dry air to get $1,7.10^{12}$ ions. Each ion has a charge of +e. What the exposure with Roentgen unit?

239. In a biological research institute, a researcher use a rabbit which has mass of 12kg to measure the whole body dose from $^{125}I_{53}$ as a radioactive source has a radioactivity of 530 Ci. When 1,2% from gamma rays arrived to this animal and the average of this rays is 0,035Mev with uniforms quantities. 60% of gamma rays reacted in the body of the rabbit. Calculate the whole body dose.

240. How much the energy is deposited in the body of a 70 kg adult exposed to a 50rad dose?

241. A dose of 500rem of gamma rays in a short period would be lethal to about half of the people subjected to it. How many rad is in this period?

242. A 70 kg person is exposed to 40mrem of alpha particles (RBE= 12). Find the absorbed dose from the biological equivalent dose (BED).

PROBLEMS AND SOLUTIONS

1. *From the comparison of the binding energy of the mirrors nuclei $^{11}B_5$ and $^{11}C_6$ estimate the value (ro) in the formula of $R = r_0 A^{1/3}$ for the nuclear radii*

Solution:

Coulomb energy is equal for the evenly charged sphere

$E = \{3/5\ e^2\}\ \{Z(Z-1)\ /\ R\}$
$\Delta E = 6e^2 Z\ /\ 5R$
$R = \{6Z/5\}.\{e^2/\Delta E\} = \{6\hbar c/\Delta E\}.\{e^2/\hbar c\} = r_0\ \sqrt[3]{A}$

Hence:

We obtain for the value of r_0:

$r_0 = 6.\ 200Mev.fm\ /\ 137.\ (11)^{1/3}.\ 3,06Mev$
$r_0 = 1,28\ fm.$

And linear sizes of the subatomic objects determine as a rule in the units of Fermi

$1fm = 10^{-13}\ cm$

We call a fine structure constant to $e^2/\hbar c = 1/137$

$\hbar c = 197.327\ Mev \cdot fm$
$\approx 200Mev.fm$
$\hbar c = 2 \cdot 10^{-11} Mev \cdot cm$

2. **Define complete E and kinetic energy E_k of an electron, the given wavelength of which equal to 10^{-2} fm. When the wavelength of the particle which expressed as λ.**

Solution:

$\lambda = \hbar / p$
  $= \hbar c / pc$
$\lambda = \hbar c (E^2 - m^2 c^2)^{-1/2}$

Hence:
When:

$\hbar = 6,582.\ 10^{-22} Mev/sec$

Or:

$h = 6,626.10^{-34} J.s$
$1ev = 1,602176.10^{-19}\ J$

Hence:

$h = 6,626068.10^{-34} J.s .\ 1(ev).\ 10^{-6} Mev$
  $1,60217610^{-19} J = 4,13566.10^{-21} Mev.sec$

And:

$\hbar = h / 2\pi$
$\hbar = 6,582.10^{-22} Mev.sec$

Hence:

$\hbar c = (6,582.10^{-22} Mev.sec).(3.10^{10})m / sec / 10^{-13}\ m / fm$
  $= 197,463 Mev.fm$
$\sqrt{\{E^2 - m^2 c^2\}} = \hbar c / \lambda = \{197,463 Mev.fm\} / 0,01 fm$
  $= 19746,3\ Mev$
  $\approx 2.10^4 Mev$

Since the rest energy of the electron m_0c^2 is only of 0.511Mev, then its total and kinetic energy practically coincide with such high energy

$$E \approx E_k \approx 2.10^4 \text{Mev}$$

These energies are accessible at present in a number of the electron accelerators of high energy

3. *Estimate the diameter of the collision of α- particle and nucleus of gold with the bombardment of a target from gold with beam of α- particles with the kinetic energy of 22Mev. Compare the result with the sum of the nuclear radii of gold and helium*

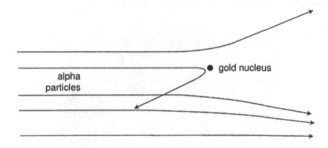

Solution:

It is a direct collision of the particle and nucleus of gold.
When the kinetic energy E_k of α- particle is expended on overcoming of the potential Coulomb barrier R then:

$$R = Z_\alpha Z_{Au} e^2 / E_k$$
$$= \{Z_\alpha Z_{Au} e^2 / E_k \, \hbar c\} \hbar c$$

And

$$E_k = Z_\alpha Z_{Au} e^2 / R$$
$$r_0 = 1,2 \text{ fm}$$

Hence

$$R_{He} + R_{Au} = r_0(4^{1/3} + 197^{1/3})$$
$$= 8,8872 fm$$
$$R_{He} + R_{Au} \approx 9 \text{ fm}$$

With the kinetic energies of α- particles is 22Mev and above distance of closest approach of the nuclei of helium and gold begins to be comparable with the sizes of nuclear systems. That means the purely Coulomb scattering reflected by the famous formula of Rutherford, does not exhaust interaction of nucleons

4. Find the density of the nuclear material.

http://institute.lanl.gov/institutes/application/

Solution:

We can find this density when we assume that the nucleus like the sphere, here we can say that the volume of this material is:

$$V = 4/3 \pi R^3$$
$$= 4/3 \pi (R_0 A^{1/3})^3$$
$$V = 4/3 R_a^3 \pi A$$

The mass of this material is:

$M_N = Zm_p + (A-Z)m_n$
$M_N = Am_p$ amu

When:

$m_p = m_n = 1$amu

The density is:

$\rho = 1{,}66.10^{-27} A / 4/3 \, \pi.(1{,}2.10^{-15})^3$
$\quad = 2{,}3.10^{17}$ kg / m³
$R = (1{,}2) \cdot \sqrt[3]{27}$
$R = 3{,}6$ fm
$R \approx 4$ fm.

5. What is the velocity and the energy corresponding for a neutron distribution in the thermal equilibrium at 30°c

Solution:

T: The temperature is in kelvin:

Hence:

$T = 30 + 273 = 303$°k
$E = \frac{1}{2} mv^2 = KT$ (By Boltzmann law)

Then:

$mv^2 = 2KT$

And:

$v^2 = (2KT/m) =>$
$v = (2KT/m)^{1/2}$
$v = (2.1,38.10^{-16}.303/1,675.10^{-24})^{1/2}$cm/sec

Hence:

v = 2234,4386 m/sec

When:

$E = KT = 1,38.10^{-16}.303$ergs$/1,602.10^{-12}$erg
E = 0,0261ev.

6. *A neutron scattered through an angle of 30° in the centre of mass by collision with a nucleus of $^{27}Al_{13.}$ Find the fractional energy which loss of this particle.*

Solution:

When:

$\alpha = (A - 1 / A + 1)^2$

And:

A is a mass number of nucleus.

Then:

$E_1 / E_0 = \{(1+ \alpha)/2\} + \{(1 - \alpha)/2\}Cos\ \theta$

Therefore:

$$E_1 / E_0 = [\{1+ (26/28)^2\} / 2]+ [\{1 - (26/28)^2\} / 2] \cos 30°$$
$$= 0{,}93112+ 0{,}068877\cos 30°$$
$$= 0{,}93112+ 0{,}05965$$
$$E_1 / E_0 = 0{,}9907$$

Therefore:

$$\Delta E = (E_0 - E_1)/E_0$$
$$= 1 - (E_1/E_0)$$
$$= 1 - 0{,}9907= 9{,}23.\ 10^{-3}$$
$$\Delta E \approx 0{,}92 \%$$

7. *A doubly ionized Helium atom is accelerated from rest through a potential difference of 6.10^8 volts in a linear accelerator.*

Calculate:
   a) *Its relativistic mass in Kg*
   b) *Its kinetic energy inMev*
   c) *Its relativistic velocity.*

Solution:

a) The moving mass is:

$$m = m_0 + dm$$
$$= m_0 + (dE_k/c^2)$$
$$m = m_0 + (qv/c^2)$$

Where:

m_0 for 4 particles is $(4.\ 1{,}67.10^{-24}$ gm$)= 6{,}68.10^{-27}$kg

And for charges it is 2e:

Hence:

q = 2. $1,6.10^{-19}$ coulomb
v = 6.10^8 joule/coulomb

Therefor:

m= $6,68.10^{-27}$+ $2.1.10^{-27}$
  = $8,78.10^{-27}$ kg

b) $E_k = (m - m_0) c^2$

When:

$$E_k = mc^2 - m_0 c^2$$
$$E_k = (m-m_0) c^2$$
$$= [(8,78 - 6,68). 10^{-27} kg.m]. (3.10^8 m/sec)^2$$

Hence:

$$= 2,1.10^{-27}. 9.10^{16} kg.m^3.sec^2$$
$$E_k = 18,9.10^{-11} kg.m^3.sec^2$$

And by the units ofMev:

$$= 18,9.10^{-11}. (1 / 1,6.10^{-13})Mev / joule$$
$$= (18,9/ 1,6).10^2 Mev$$
$$E_k = 1180 Mev$$

c) Rather than the substitute values of m_0 and m in the equation:

$$m= m_0 / \sqrt{\{1- (v/c)^2\}}$$

When:

$$\theta = v / c$$

Then we have:

$$\sqrt{\{1-(v/c)^2\}} = \sqrt{(1-\sin^2\theta)} = \cos\theta$$

The mass variation equation becomes:

$$\cos\theta = m_0/m$$

Hence:

$$\cos\theta = m_0/m$$
$$= (6,68.10^{-27} / 8,78.10^{-27})$$
$$= 0,76$$
$$\theta = 40°30 \text{ and } \sin\theta = 0,65$$

Therefore:

We get $\sin\theta = v/c = 0,65$

$$v = 0,65 . c = 0,65 . 3.10^8 \text{ m/sec}$$
$$v = 1,95.10^8 \text{ m/sec.}$$

8. ***Compare the given wavelengths of electron and proton with the identical kinetic energy of 100Mev***

Solution:

For the electrons

$$\lambda_e = \hbar c/(E_k^2 + 2E_k . m_e c^2)^{1/2}$$
$$= 200 \text{Mev.fm}/(10^4 + 2.10^2.0,511)^{1/2}\text{Mev}$$
$$\lambda_e \approx 2\text{fm}$$

For the protons:

$\lambda_p = \hbar c/(E_k{}^2 + 2E_k . m_p c^2)^{1/2}$
$\quad = 200 Mev.fm/(10^4 + 2.10^2 .938)^{1/2} Mev$
$\lambda_p \approx 0.45 fm$

9. Determine the complete energy (E) and the kinetic energy (E_k) of an electron when the wavelength of this electron is equal to 0,01fm

Solution:

The given wavelength of the particle is expressed as

$\lambda = \hbar/p = \hbar c/pc = \hbar c(E^2 - m^2 c^4)^{-1/2}$

From where:

$\{\sqrt{E^2 - m^2 c^4}\} = (\hbar c/\lambda)$
$\qquad = 200 Mev.fm/0,01fm$
$[\sqrt{E^2 - m^2 c^4}] = 2.10^4 Mev$ (See problem 2 about this energy)

10. Calculate the wavelength of electrons with 35Gev as energy

Solution:

When the 35Gev= 35000MeV= $3,5.10^{10}$ev

And the kinetic energy of an electron is

$E_K = mc^2 - m_0 c^2 \approx mc^2$

And

$$E^2 = P^2c^2 + M_0^2c^4 \approx P^2c^2$$

And when the rest energy is small with comparison with Pc

Hence

$$P^2c^2 = m^2v^2c^2 \approx m^2 c^4$$

Hence

$$v = c$$

When

$$\lambda = h/m\,v$$
$$\approx h/mc$$
$$\lambda = hc/mc^2$$

Hence

$$\lambda = (6,6.10^{-34} \text{ J.s})(3.10^8 \text{ m/s})/(3,5.10^{10}\text{eV})(1,6.10^{-19}\text{J/eV})$$
$$= 3,535.10^{-17}\text{m}$$
$$\lambda = 0,035\text{fm}$$

This result is small with comparison with the volume of nucleus.

11. Calculate the radius and the approximate mass of the nucleus of $^{40}Ca_{20}$. And find the approximate density of this nucleus also.

Solution:

The radius equation of the nucleus is:

$$R = R_0 A^{1/3}$$

When R_0 is an experimental constant, equal $1,2.10^{-15}$m= 1,2 fm

Therefore:

$R = 1,2.10^{-15}$ m. $(40)^{1/3}$
$R = 4,104.10^{-15}$m

With the negligence of the mass difference between the proton and the neutron, we can say that the mass of this proton (or neutron) is $1,67.10^{-27}$kg.

Hence:
The mass of all nuclides in the nucleus M is:

M= 40. $1,67.10^{-27}$= $6,68.10^{-26}$ kg.

With negligence of the nucleus binding energy in the calculation we get:

$V= 4/3 \pi R^3$= $4/3 \pi (4,104.10^{-15})^3$
$V= 2,894.10^{-43}$m^3

Therefore:
The density is:

$\rho = m / v$= $6,68.10^{-26}/2,894.10^{-43}$
$\rho = 2,308.10^{17}$kg/m^3

12. to what is equal velocity of particle v, whose kinetic energy T is equal to its rest energy mc^2?

Solution:

The total energy of the relativistic particle is:

$E= E_k + mc^2 = mc^2 / \sqrt{(1- v^2 / c^2)}$

In the case of $E_k = mc^2$ we obtain

$2mc^2 = mc^2 / \sqrt{(1- v^2 / c^2)}$

Hence:

$v = (\sqrt{3}) / 2\ c$
$v \approx 0,87\ c$

13. The kinetic energy of a neutron is 4,8Mev collide the nucleus of $^4He_2$ which will recoil if struck head-on calculate the kinetic energy of $^4He_2$.

Solution:

By the two laws of the conservation of the energy and of the momentum, when the masses of the nucleus of helium and of the neutron are M and m respectively.

$\frac{1}{2} mv_0^2 = \frac{1}{2}(MV^2) + \frac{1}{2} mv^2$
$\quad mv_0 = MV + mv$ \hfill (*)

From these equations we can write the following form:

$v_0 + v = V$ \hfill (**)

From (*) and (**) we obtain:

$v = - v_0 \{(M-m) / (M+ m)\}$

From this form we get the following form:

$E / E0 = \{(M-m) / (M+ m)\}^2$

And:

$\Delta E = E0 - E = [4mM / (M+ m)2].$ E0

From this relation we get the result of the kinetic energy of Helium nucleus:

$\Delta E = E0 - E = [4mM / (M+ m)^2].$ E0
$\Delta E = E0 - E = [4. 1. 4 / (4+ 1)^2].$ 4,8
$\Delta E = 3,072Mev.$

14. An alpha particle has 6,5Mev as a kinetic energy, approach from a foil of gold. Calculate the impact parameter and the cross section in this collision when θ= 45⁰, and calculate the distance of the smallest approach in this collision.

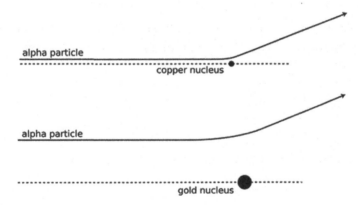

Solution:

When the velocity of this particle is a few relative with the velocity of light:

Hence:

$E_k = 6,5.10^6. 1,6.10^{-19}$ J
$E_k = 1,04.10^{-12}$ joule.
$\Delta m = E_k / c^2 = 1,04.10^{-12} / (3.10^8)^2$
$\Delta m = 1,155.10^{-29}$ kg

When the mass of an alpha particle is:

$m_{\alpha 0} = 6,645.10^{-27}$ kg
$M_\alpha = m_{\alpha 0} + \Delta m$
$\quad = 6,645.10^{-27}$ kg$+ 1,155.10^{-29}$ kg
$M_\alpha = 6,656.10^{-27}$ kg
$M_\alpha = m_{\alpha 0} / \sqrt{\{1-(v^2/c^2)\}}$
$(M_\alpha / m_{\alpha 0})^2 = 1 / \{1-(v^2/c^2)\}$

Hence:

$v = 0,063$ c

The impact of collision is b:

$b = \cot (45/2).\ 9.10^9.79.(1,6)^2.10^{-38} / 1,04.10^{-12}$joule

Hence:

$b = 1,75.10^{-14}$ m

The cross section in this collision σ:

$\sigma = \pi\, b^2 = 3,14\ (1,75.10^{-14})^2$
$\sigma = 9,616.10^{-28}$ m^2

The distance of the smallest approach in this collision D:

$D = 2kZe^2 / E_k$
$\quad = \{2(9.\ 10^9)(79)(1,6.10^{-19})^2\} / \{1,04.10^{-12}\}$
$D = 3,5.10^{-14}$ m

15. A particle with 20Mev as a kinetic energy moves in a curvature, find the radius of this curvature when the movement of this particle is perpendicular on a magnetic field of 0,1 W/m^2

Solution:

We know that from the nonrelativistic mechanics, the charge q of the particle and with p momentum, moves perpendicular on the inducted uniform magnetic field B, and this movement will be with the curvature which has R radius when:

P= BqR

When the velocity of the particle nears the velocity of the light, so the suitable equation is the relativistic equation:

$$P = mv = m_0 v / \sqrt{ 1-(v/c)^2 }$$

When the total energy E is:

$$E = (E_k + m_0 c^2)$$
$$E = 20,511 Mev.$$

When:

$$E^2 = p^2 c^2 + m_0^2 c^4$$

Hence:

$$P^2 = (1/c^2). \{ E^2 - (m_0 c^2)^2 \}. (1,6.10^{-6} erg)^2$$

Hence:

$$P = 10,9.10^{-14} erg. \ sec/cm$$
$$P = 10,9.10^{-21} joule.sec /cm$$

When:

$$q = 1,6.10^{-19} C , \ and \ B = 0,1 \ weber / m^2$$

Hence:

R = P/qB= 10,9.10^{-21} / {1,6.10^{-9} coulomb)(0,1 weber/m^2)
 = {10,9 / 0,16 }. 10^{-2}m
R = 0,68m = 68cm.

16. What is the sum of the kinetic energy of the production of gamma ray, when the energy of this ray is 12Mev?

Solution:

The energy of gamma ray is 12Mev , that means:

E_{k1}(for an electron)+ E_{k2}(for a positron)=
  = energy of γ ray − 2(the rest energy of β⁻ or β⁺)
E_{k1}+ E_{k2} = hv − 2(E$_0$)

When:

hv = 12Mev , and 2E$_0$= 1,022Mev:

Therefore:

E_{k1}+ E_{k2} = 12 − 1,022
E_{k1}+ E_{k2} = 11,022Mev.

17. A cyclotron in a research laboratory has a radius of R= 0,5m. When the magnetic field in this cyclotron is 2,3T.

a) Find the frequency in this case.
b) Calculate the kinetic energy of the accelerated protons when they leave this accelerator.

Solution:

a) When:

$f = 1/T$, f is the frequency and T is the periodic time.
$T = 2\pi r / (qBr/m)$
$T = 2\pi m / qB$

When r is the radius; q is the proton charge which equal to the charge of the electron; B is the magnetic field intensity; m is the mass of the accelerated proton.

Hence:

$f = qB / 2\pi m$
$= (1,6.10^{-19}).(2,3) / (2. \, 3,14. \, 1,67.10^{-27})$
$f = 2,6.10^{7}$ Hz.

b) When the proton which leaves has the radius which the same radius of the cyclotron:

Hence:

r= R= 0,5 m

When:

$v = qBr / m$

Therefore:

$E_k = \frac{1}{2} mv^2 = \frac{1}{2} m(q^2B^2R^2 /m^2) = (q^2B^2R^2 / 2m)$
$= (1,6.10^{-19})2.(2,3)2.(0,5)2 / 2.1,67.10^{-27})$
$E_k = 1,014.10^{-11}$ joule.

18. *What the distance of the intensity of the muons beam with a kinetic energy of E_k = 0.5Gev which move in the vacuum, decreases to half of the initial of magnitude?*

Solution:

Reduction of the intensity of the beam of muons occurs as a result of the disintegration of muons:

$$\mu^- \rightarrow e^- + \acute{\upsilon}_e + v_{\mu^-}$$

The number of muons N(t) which were not decomposed up to the moment of the time t. It is determined by the relationship of:

$$N(t) = N(0)\, e^{(-t/\tau)}$$

Where τ- the average life of muon; N(0) – the number of muons at the initial moment of the time. The average life of the quiescent muons is equal of

$2.2 \cdot 10^{-6}$s. In this case:

$$N(t) = N(0)/2 = N(0)\, e^{(-t/\tau)}$$

I.e. exp(- t/τ)= 1/2, or t= τln2. The relativistic retarding of the course of time is determined by the relationship of

$$t = t_0 / \sqrt{(1-(v^2/c^2))}$$

Where t_0–the time in the system connected with the moving body. In our case

We obtain:

$$t = \tau \ln 2\ / \sqrt{(1-(v^2/c^2))} \qquad \qquad ¤$$

Connection between the kinetic energy E_k and pulse p of the particle

$P = [\sqrt{(E_k^{2+} 2E_k mc^2)}] / c$ #

Relativistic pulse of the particle

$P = mv / \sqrt{(1-(v^2/c^2))}$ ##

Where m – the rest mass of the particle; v - its speed. We will obtain from # and ##

$v = [p. \sqrt{(1-(v^2/c^2))}] / m$ @@
$\{\sqrt{(E_k^{2+} 2 E_k mc^2)}\}. [\sqrt{(1-(v^2/c^2))}] / mc$

The rest energy mc^2 of muon is 10^6Mev. The path of muon

$l = vt$ #¤

Substituting in #¤, ¤ and @@, we will obtain

$l = \{\tau c \ln 2. [\sqrt{(E_k^2 + 2E_k mc^2)}]\} / mc^2$

$\{2,2. 10^{-6}$ s. $3,10^{10}$cm/s . $0,693. \sqrt{\{(500Mev)^2 + 2.500Mev . 10^6 Mev\}}]$ / 10^6Mev

$l = 2,6 . 10^5$cm

19. Calculate the wavelengths λ of a proton and of an electron with the kinetic energy E_k of 10Mev

Solution:

Proton nonrelativistic ($E_{kp} \ll m_p c^2$). In this case:

$\lambda = \hbar/p \approx \hbar c / (\sqrt{2mc^2 E_k})$

Taking into account that $\hbar c$= 197 Mev·fm, we have

λ_p = (197Mev. fm) / ($\sqrt{}$ 2. 938,3Mev . 10Mev

$\lambda_p \approx 1,4$ fm

Electron is relativistic ($E_{k\,e} >> m_e c^2$). In this case:

$\lambda_e = \hbar c/ E_{ke}$

   = (197Mev.fm) / 10Mev

$\lambda_e \approx 20$fm

20. A proton, electron and photon have identical (l) wavelength

$\lambda = 10^{-9}cm$. *What the time t by them necessarily for the flight path of L= 10 m?*

Solution:

For the proton and for the electron

$p= \hbar/\lambda$

$p= mv / \sqrt{(1-(v^2/c^2))}$

T.e. $\hbar/\lambda= mv / \sqrt{(1-(v^2/c^2))}$

Hence:

We get that:

$V= (\hbar c^2)/ \{\sqrt{(mc^2)^2\lambda^2+ \hbar^2 c^2}\}$

The time of the stairwell:

$t= L/v= \{L [\sqrt{(mc^2)^2\lambda^2+ (\hbar c)^2}]\} / (\hbar c.c)$

Dr. Mouaiyad M.S.Alabed

Proton

$t = 10^3 cm[\sqrt{(938,3Mev)^2. (10^4fm)^2 + (197Mev. fm)^2}] / (197Mev.fm. 3.10^{10}cm/s)$

$t \approx 1,6.10^{-3}s$

Electron

$t = 10^3 cm\sqrt{[(0,511Mev)^2. (10^4fm)^2 + (197Mev. fm)^2]} /(197Mev.fm. 3.10^{10}cm/s)$

$t \approx 0,9.10^{-6}s$

Foton

$t = L/c = 10^3 cm / 3.10^{10}cm/s$
$t \approx 3,3. 10^{-8}s$

21. The wavelength of a photon is $\lambda= 3 \cdot 10^{-11} cm$. Calculate the momentum p of this photon

Solution:

$P = \hbar/\lambda$
$= \hbar c /\lambda c$
$\approx (197Mev.fm)/300fm.s$
$P \approx 0,66Mev/s$

22. Estimate radii of the atomic nuclei $^{27}Al_{13}$, $^{90}Zr_{40}$ and $^{238}U_{92}$

Solution:

The empirical dependence of the nuclear radius R on the number of nucleons A (A > 10), $R \approx r_0 A^{1/3}$. The parameter $r_0 \approx 1,23 \cdot 10^{-13}$ cm $\approx 1,23$fm is approximately identical for all nuclei n by the relation of the radius:

50

$R \approx r_0 A^{1/3}$

For: $R_{27Al} = 1,23fm. (27)^{1/3} = 3,7fm$

For: $R_{90Zr} = 1,23fm. (90)^{1/3} = 5,5fm$

For: $R_{238U} = 1,23fm. (238)^{1/3} = 7,6fm$

23. Estimate the density of a nuclear material when the mass of one nucleon in nucleus is $m_N \approx 1amu = 1.66 \cdot 10^{-24} g$

Solution:

The density of the nuclear material is the nuclear mass divided into its volume

$\rho = (m_N A) / (4/3) \pi R^3$

$= (3m_N A)/(4\pi r_0^3 A)$

$= 3m_N/4\pi r_0^3$

$= 3.1,66.10^{-24}g/4. 3,14.(1,3. 10^{-13}cm)^3$

$= 1,8.10^{14}g/cm^3$

$\rho = 180$ mln. Ton/cm^3

The density of the nuclear material does not depend on A

24. Calculate the rest energy for a material of 2,5gram.

Solution:

When:

$E_0 = mc^2$

$E_0 = \{2,5.10^{-3} (3.10^8)^2\}J$

Therefor:

$E_0 = 2,25.10^{14}J.$

25. The nucleus of $^{60}Co_{27}$ irradiate a photon with the energy of E_γ= 2,15.10⁻¹³J How many mass defect is in this isotope?

Solution:

When the cobalt nucleus irradiate gamma rays then the energy of this mass will decrease by the value of Δm and $\Delta E_0 = E_\gamma = 1,13.10^{-13}J$

Then:

We express about this relation as the form of:

$\Delta E_0 = \Delta m.c^2$

Therefor:

$\Delta m = \Delta E_0 / c^2$
     $= \{2,13.10^{-13}/(3.10^8)^2\}Kg$
$\Delta m = 2,37.10^{-30}$ Kg

This value equal to three times of the mass of the electron approximately.

26. Estimate the nuclear radius of $^{40}Ca_{20}$ with an elastic scattering of electrons when the kinetic energy E_k is 750Mev on the nuclei of $^{40}Ca_{20}$ in the section which observed the diffraction minimum at angle θ_{min}= 18°

Solution:

The situation of the first minimum in the elastic scattering cross section θ_{min} is possible to estimate with the aid of the formula for diffracting the plane wave on the disk of the radius R

$\sin \theta_{min} = 0,6 \lambda/R$

The taking into account that electrons ultra-relativistic. We obtain

$R = 0{,}6\lambda/\sin\theta_{min}$
$= [0{,}6 / \sin\theta_{min}][2\pi\hbar c/E_k]$
$= 0{,}6.6{,}28.197 Mev.fm / 750 Mev.0{,}31$
$R \approx 3{,}2$ fm

27. Calculate the velocity of an electron in the rest state which has been accelerated through a potential difference of 850kv

Solution:

By the relation of the kinetic energy:

$K = e\,\Delta v = 850 kev$

When the rest energy of an electron is:

$E_{oe} = m_o c^2 = (9{,}1.\,10^{-28} gm)\,(3.10^{10} cm/sec)^2$
$= 8{,}19.10^{-7} erg$
$E_{oe} = 511$ kev

The total energy (E) is:

$E = m_o c^2 + E_k$
$= 511 + 850 = 1361$ kev

Therefore:

$E / m_o c^2 = 1 + E_k / m_o c^2$

Or:

$mc^2 / m_o c^2 = 1 + (850 / 511)$
$= (511 + 850)/511$
$mc^2 / m_o c^2 = 2{,}6634$

Hence:

$m = 2,6634m_0$
$m = m_0(1 - v^2/c^2)$

Therefore:

$1 - v^2/c^2 = 1/(2,6634)^2$
$v/c = (1 - 1/7,0937)^{1/2}$
$\quad = (0,859)^{1/2}$
$v/c = 0,9268$

Hence:

$V_e = 0,9268\ C$
$V_e = 2,78.10^{10}cm/sec$

28. A mass of a neutral atom of $^{16}O_8$ is 15,9949amu and for $^{15}O_8$ is 15,0030amu and for $^{15}N_7$ is 15,0001amu What are the energy of the branches of the neutron and of the proton in $^{16}O_8$ nucleus?

Solution:

The energy of the branches of the neutron:

$\varepsilon_n(A,Z) = m_n + m(A-1,Z) - m(A,Z),$

Of the proton:

$\varepsilon_p(A,Z) = m_p + m(A-1,Z-1) - m(A,Z)$

In both formulas the masses must be in the energy unit for the nucleus of ^{16}O

$\varepsilon_n = 939,57144\ Mev + (15.0030u. - 15.9949u.)\ 931,5Mev$
$\varepsilon_n = 15,60262Mev$

$\varepsilon_p = 938,78898\text{Mev} + (15.0001\text{amu} - 15.9949\text{amu}) \cdot 931,5\text{Mev}$
$\varepsilon_p = 12,11881\text{Mev}$

29. An isotope of $^{191}Os_{76}$ has a radioactivity of 2,4 curie uses in a purpose for 8,65 days. What is the activity of this isotope at that time?

Solution:

The decay constant of this isotope is:

$\lambda = 0,693/T_{1/2}$
$= 0,693/15,4.24.3600$
$= 0,693/1330560$
$\lambda = 5,208. \ 10^{-7} \ \text{sec}^{-1}$

The activity in this case is:

$2,4.3,7.10^{10} = 8,88.10^{10} \ \text{dis/sec}$

And:

$A = \lambda N_0$
$= 5,208.10^{-7}.N_0$
$A = 8,88.10^{10}$

And:

$\lambda N = \lambda N_0 \ e^{-\lambda t}$
$= 8,88.10^{10.}$
$= 8,88.10^{10}.e^{-5,208.10-7.(8,65.24.3600)}$
$= 8,88.10^{10}.e^{-0,389225}$
$= 8,88.10^{10}.0,6775819$
$\lambda N = 6,0169. \ 10^{10} \ \text{dis/sec}$

30. What is the minimum energy T_{min} for a deuteron which must have in order as a result of an inelastic scattering on the nucleus of $^{10}B_s$ to excite state with the energy E_{ex} of 1.75Mev?

Solution:

With the inelastic scattering the reaction energy $Q = -E_{ex}$ and the minimum energy of the deuteron equal to the threshold of the reaction $E_{kmin} = E_{kth}$

After using of the formula for the threshold of the reaction (since $Q << m_d c^2$)

We obtain:

$$E_{kmin} \approx E_{kth} (1+m_1/m_2)$$

When:

$m_d = 2$ and $m_B = 10$

Hence:

$$E_{kmin} \approx 1.75 \ (1+2/10)$$
$$\approx 1,75 \ (1,2)$$
$$E_{kmin} \approx 2.1 \ Mev$$

31. Compare the mass of a $^{40}Ca_{20}$ nucleus to that of its constituent nucleons.

Solution:

The mass of a neutral $^{40}Ca_{20}$ is 39.962591amu and the mass of the neutrons and of the protons are:

$20 \ m_p = 20 \ (1.007825) = 20.1565amu$
$20 \ m_n = 20 \ (1.008665) = 20.1733amu$
$20 \ m_p + 20 \ m_n = 40.3298amu$

40.3298amu - 39.962591amu= 0.367209amu
The difference between them is: 0.367209amu

32. Can we get β^- in the decay process of $^7Be_4$ if the daughter nucleus is $^7Li_3$?

Solution:

The mother nucleus is $^7Be_4$.

And:

M_{Be}= 7,016928amu, M_{Li}= 7,016003amu

Therefore:

ΔM = 7,016928 - 7,016003= 9,25.10⁻⁴amu

And:

Q= 9,25.10⁻⁴amu.931,5Mev/amu
Q= 0,8616375Mev

This energy is less than the energy equivalent of an electron and of an positron (which they are same!) masses which is 1,022Mev.

33. π meson has a mass of 135Mev/C². What is the mass in atomic mass units (amu)?

Solution:

m_π= 135Mev/ C²
And 1amu= 931.5Mev/C²
1amu of π= 135/931.5= 0.145amu

The various scattering experiments performed indicate the approximate radius of nuclei of the different elements. The radius r is found to depend on the mass which in turn depends on the mass number A. It is found that the radius is approximately given by:

$$R = r_o . A^{1/3}$$

Where r_o is the same constant for nuclei of all elements.
The volume V, being proportional to r^3 is given by:

$$V \alpha r^3$$

Hence:

$$V \alpha (r_0 A^{1/3})^3$$

And:

$$V \alpha A$$

$$V / A = constant$$

This shows that the density of the nuclei is more or less than constant. The approximate radius of the nucleus depends on the atomic mass number A

34. a) What is the approximate radius of $^{64}Cu_{29}$ nucleus?
b) Approximately what is the value of A of a nucleus which has a radius of $3,6.10^{-15}$ m?

Solution:

a) $r \approx (1,2. 10^{-15}$ m$). A^{1/3}$
    $\approx (1,2. 10^{-15}$ m $). (64)^{1/3}$
    $R \approx 4, 8. 10^{-15}$ m

b) r= constant $A^{1/3}$

$3,6.\ 10^{-15} \approx 1,2.\ 10^{-15}.\ A^{1/3}$

$\therefore A^{1/3} = 3,6.\ 10^{-15} / 1,2.\ 10^{-15} = 3$

And $1/3 \ln A = \ln 3$

And $\ln A = \ln 3 / 0.3333 = 3.2958$

$\therefore A \approx 27$

The nucleus is Aluminum $^{27}Al_{13}$.

35. A neutron with an energy of 6Mev collide a $^{239}Pu_{92}$ nucleus. Find the velocity of $^{239}Pu_{92}$ nucleus thrown forward directly.

Solution:

When m_1 , v_1 , m_2 , $v`_2$ and $v`_1$ are respectively: the mass and velocity of the neutron before collision, the mass and velocity of the struck particle after collision, and $v`_1$ is the velocity of the incident particle after collision, then:

$½ m_1 v_1^2 = ½ m_1 v`_1^2 + ½ m_2 v`_2^2$

$m_1 v_1 = m_1 v`_1 + m_2 v`_2$

We get from these equations:

$v`_2 = \{2m_1 / (m_1 + m_2)\}.\ v_1 = \{(2.\ 1) / (1 + 239)\}.\ v_1$

But $v_1 = \sqrt{\{(2.\ 6.\ 1,6.\ 10^{-6}) / (1,67.\ 10^{-24})\}} = \sqrt{1,1497.10^{19}}$cm/sec

$= 3,39.\ 10^9$cm/sec

Therefore:

$v_2 = (2 / 240).3,39.10^9$

$= 28256027$

$v_2 = 0,2825.10^9$cm/sec.

36. What is the approximate mass of Neptunium $^{239}Np_{93}$ nucleus?

Solution:

Approximate mass (ingrams)= atomic mass / Avogadro`s number
$$= 239,052932/6,025. \, 10^{23}$$
$$= 3,968.10^{-22} \text{grams}$$

37. How much energy is required to remove a neutron from $^{236}U_{92}$ nucleus?

Solution:

The mass of $^{236}U_{92}$ is 236,045562amu and the remaining nucleus is $^{235}U_{92}$, its mass is 235,043924amu

The mass of $^{1}n_{0}$ is 1,008665amu Therefore the mass of the final system is:

235,043924+ 1,008665= 236,05258amu

Hence:
The mass increase of the system is:

236,05258 - 236,045562 = 7,02.10^{-3}amu

Thus the energy required to remove a neutron from the nucleus of $^{236}U_{92}$ is:

7,02.10^{-3}. 931,5= 6,54Mev.

38. What is the stable nucleus which has a radius equal to the radius of the uranium nucleus?

Solution:

$r \approx (1,2. \, 10^{-15}m).A^{1/3}$

To the uranium nucleus is:

$r_u \approx (1,2. \ 10^{-15}m).A^{1/3}. \ (238)^{1/3}$
    $= 7,4365848. \ 10^{-15}m$
$r_{nuc} = 3,7182858.10^{-15}m$

When:

$r \approx (1,2. \ 10^{-15}m).A^{1/3}$

Hence:

$3,7182858.10^{-15}m \ / \ (1,2. \ 10^{-15}m) = A^{1/3}$
        $= 3,0985715 = A^{1/3}$

Hence:

$A \approx 29,749 \approx 31$

And:

$^{31}X_{15}$ is $^{31}P_{15}$ Phosphorous.

39. The $^{214}Po_{84}$ nucleus undergoes an alpha decay. What will be the alpha kinetic energy?

Solution:

We assume that the mother nucleus is at rest before decay, but the daughter nucleus recoil, carrying with it a certain amount of the energy. Alpha disintegration energy is the sum of the two products: alpha and the daughter nucleus. From the conservation of the momentum:

$mv = m_{po} \ V_{po}$

The energy of alpha decay (E_α) is:

$E_\alpha = 1/2mv^2 + 1/2\, m_{po}\, v_{po}^{\,2}$
$\quad = 1/2mv^2\, \{1+(\,m/m_{po})\} \equiv$ mass of parent nucleus / mass of product nucleus
$\quad = A\,/\,(A-4)$
$E_\alpha = 214\,/\,(214-4)$
$\quad = 214\,/\,210$
$E_\alpha = 1,0190476$ Mev.

40. A proton accelerated in a ring acceleration which has 2T as a magnetic field. Find the periodic frequency when the mass of the proton is $1,67.10^{-27}$ kg.

Solution:

Hence:

$r = m_p\, v\,/\,Bq$
m_p: mass of the proton; v: velocity of the proton; q: charge of the proton
$\quad$; B: intensity of the magnetic field; r: the radius and when
$v = w\,/\,2\pi = v\,/\,2\pi\, r$
v: periodic frequency; w: angular velocity of the proton(when it is rotate in the accelerator)

Hence:

$v = v\,/\,2\pi\,(mp\, v\,/\,qB) = Bq\,/\,2\pi\, mp$
$\quad = 2(1,602177.\,10^{-19})\,/\,2.\,3,14\,(1,67262.10^{-27})$
$v = 30,49$ MHz

41. The $^{232}U_{92}$ nucleus irradiates α- particle with the kinetic energy of $E_{k\alpha}$ of 5,32Mev. Find the daughter nucleus, which the

value of the approximate atomic mass (withamu units) for the daughter atom.

Solution:

The reaction is:

$$^{232}U_{92} \rightarrow {}^{228}X_{90} + {}^{4}He_{2}$$

The type of the daughter nucleus $^{228}X_{90}$ is $^{228}Th_{90}$ which follow an alpha particle and gamma quant and it has a half-life of 1,9131year.

Therefor:

To find the approximate atomic mass:

$$M_{Th90} = M_{U92} - M_{\alpha}$$
$$= 232,037131amu - 4,002602amu$$
$$M_{Th90} = 228,03453amu \text{ it's the daughter nucleus.}$$

The final atom is by adding the approximate mass of the electrons which is

90e⁻.

Therefor:

$m_{e^-} = 9,1093897. 10^{-31}$ kg.

When: 1amu= $1,6605.10^{-27}$ kg.

Hence:

$m_{e^-} = 5,4859317.10^{-4}$amu

And:

90e⁻= 0,0493733amu

Hence:

M_{Th}= 228,03453amu+ 0,0493733amu
M_{Th}= 228,0839amu

42. When $^{23}Ne_{10}$ (mass = 22,9945amu) decay to $^{23}Na_{11}$ (mass = 22,9898amu), what is the maximum kinetic energy of the emitted electron? What are the minimum energy and the energy of the neutrino in each case?

Solution:

$^{23}Ne_{10}$ $\longrightarrow$ $^{23}Na_{11}+ \beta^{-}+ \acute{\upsilon}$

We find Δm:

$\Delta m = 22,9945 - (22,9898+ m_{e^-}+ m_{v})$

When: $m_{e^-}= 0,0047$

Hence:

$Q_{max}= 0,0047. 931,5 = 4,37805$Mev (maximum K.E. of the emitted electron).

43. Proof how we can express the atomic mass unit for?

Solution:

When: 1amu= 1,66054. 10^{-27} kg

The Einstein relation is:

$E = mc^2$

Consequently $E = (1,66054. \ 10^{-27} \text{ kg}) \ (2,997924. \ 10^8 \text{ m/s})^2$

$= 1,4924188. \ 10^{-10} \text{ J}$

$= 1,4924188. \ 10^{-10} \text{ J} \ / \ 1,6022. \ 10^{-13} \text{ J/Mev}$

$E = 931,48 \text{Mev}.$

This value is energy withMev unit, and this express for any energy for any nuclear reaction. When we want use this unit instead ofamu, where this unit equal (1amu= 1,66054. 10^{-27}kg) from the interactive mass or this mass which result from this reaction.

44. Compare between the mass of helium ($^4He_2$) nucleus and a nucleus which has some nucleons.

Solution:

The mass of the neutral helium atom $^4He_2$ from the table:

$M_{4He2} = 4,002602$amu

The mass of two neutrons and of two protons (which included two electrons) is:

$2m_n = 2,01733$amu

$2m_{(1H1)} = 2,01565$amu

We calculate the nucleon mass:

$2m_n + 2m_{(1H1)} = 2,01733amu+ 2,01565$amu

$= 4,032980$amu

Hence the difference between the nucleus of helium and the nucleus is:

$4,032980 - 4,002602 =$

$= 0,030378$amu

45. A mass with 10grams. Calculate its energy with ev unit, which fall from the origin under effect of gravitational force, with the distance of 15cm. Compare this result with fall hydrogen molecule energy through the same distance.

Solution:

From the linear motion equations:

$V^2 = V_0^2 + 2ax$

When:

V: the final velocity for falling mass.
V_0: the initial velocity for falling mass.
a: mass's acceleration (when a= g)
x: the mass displacement.

And when $V_0 = 0$, a= g

Hence:

$V = \sqrt{(2gx)}$
  $= \sqrt{2.980.\ 15}$
  $= 171,46cm\ /\ sec$
$V = 1,7146\ m\ /\ sec.$

When:

$E_K = \frac{1}{2}\ mv^2$

Hence:

$E_K = \frac{1}{2}\ .\ 10grams.\ (171,46)^2$
  $= 146992,65ergs.$

And:

1 ev = 1,6. 10^{-12}ergs.

Hence:

E_K= 146992,65erg / 1,6. 10^{-12}erg / ev.

For this molecule, when the proton mass is 1,67252. 10^{-24}gram

Hence:

For the two protons= 3,345. 10^{-24} gm.

And:

E_{KH}= 4,9169. 10^{-20}

Hence:

E_K = 3,073. 10^{-8} ev.

46. What is the approximate mass of Neptunium $^{239}Np_{93}$?

Solution:

Approximate mass (in grams)= Atomic mass / Avogadro's number

= 239,052932 / 6,025. 10^{23}
= 3,968.10^{-22}grams.

47. a)   What is the approximate radius of $^{64}Cu_{29}$ nucleus?
   b)   Approximately what is the value of A for a nucleus whose radius is $3,6.10^{-15}$ m?

Solution:
   a) $r \approx (1,2.10^{-15}).\ A^{1/3}$
          $\approx (1,2.\ 10^{-15}).\ (64)^{1/3}$
          $\approx 4,8.\ 10^{-15}$ m
   b) $r = $ constant $A^{1/3}$
          $3,6.10^{-15} \approx 1,2.10^{-15}.\ A^{1/3}$

Hence:

$A^{1/3} = (3,6.10^{-15}) / (1,2.\ 10^{-15}) = 3$
$(1/3) \ln A = \ln 3$

Hence:

$\ln A = \ln 3/(1/3) = 3,2958$
$A \approx 27$
So the nucleus is Aluminum:$^{27}Al_{13}$

48. How long will it take a 1Mev neutron to cross a $^{238}U_{92}$ nucleus?

Solution:

When: $R = r_0\ A^{1/3}$
Hence:

$2R = 2.\ 1,25.10^{-13}\ (238)^{1/3}$
      $= 1,549.10^{-12}$cm

When the velocity of 1Mev is $1,4.10^9$cm/sec , therefore the time required for this energy to transverse the nucleus is:

$= 1,549.10^{-12} / 1,4.10^9 = 1,1066.10^{-21}$sec

49. The velocity of alpha particles from plutonium $^{239}Pu_{94}$ is $1,5.10^7$ m/sec. What is the velocity of the recoil atom?

Solution:

$M_{pu} V_{pu} = M_U V_U - m_\alpha v_\alpha$

When: $V_{pu} = 0$, V_U and v_α have opposite directions,

Therefore:

$M_U V_U - m_\alpha v_\alpha = 0$
$M_U V_U = m_\alpha v_\alpha$

And: $V_U = (m_\alpha / M_U). v_\alpha$

But: $(m_\alpha / M_U) = 4 / (A-4)$
$V_U = (4 / (A-4)). v_\alpha$
$V_U = (4 / (239-4)). 1,5.10^7$ m/sec
$\quad = 0,0170212. 1,5.10^7$ m/sec
$\quad = 255319,14$ m/sec

Hence:

$V_U = 2,55. 10^7$ m/sec.

50. What is the fractional increase in mass of a 10Mev proton?

Solution:

The total energy of proton is E as:

$E = E_K + E_0$

$E_K = E + E_0$
$\quad = mc^2 - m_o c^2$
$E_K = c^2 (m - m_0)$

69

When: $E_K = 10 Mev.$
So: $E_K = 10.10^6 . 1,6.10^{-12} erg$
$E_K = 1,6.10^{-5} erg$

Therefore:

$m - m_0 = E_K / c^2 = 1,6.10^{-5} erg / (3.10^{10} cm/sec)^2$
$\approx 1,78.10^{-26}$ gm$= 1,78.10^{-29}$ kg

The fractional increase in this mass:

$\Delta m / m_0 = 1,78.10^{-26} . 10^{-3}$ kg $/ 1,672623.10^{-27}$ kg

When $m_{0p} = 1,672623.10^{-27}$ kg

Therefore:

$\Delta m / m_0 = 0,0106$
$\Delta m / m_0 \approx 0,01$

51. What is the neutron temperature corresponds to 8ev?

Solution:

$E = 1/2 \, mv^2 = KT$, by Boltzmann's law

When K is Boltzmann's constant$= 1,380658.10^{-23}$ J/k

Or: $K = 1,380658.10^{-16} erg/^0 k$.
And T= temperature in degrees Kelvin(^{0}k).

From the Boltzmann's law:

$T = E_K / K = 8 . 1,6.10^{-12} / 1,380658.10^{-16}$
$= 9270942^0 k$
$T = 9,27.10^4 \, ^0 k$

52. Estimate the value of the charge number Z, with which the nuclei become unstable with the relation to the spontaneous decay

Solution:

Spontaneous nuclear decay appears in the case of the Coulomb repulsion of the nucleus protons begins to predominate above the nuclear forces tightening nucleus. The estimation of the nuclear parameters, with which sets in this situation, can be carry out from the consideration of changes in the surface and Coulomb energies with the nuclear distortion. If the deformation leads to the more advantageous energetically state, nucleus will be spontaneously deformed up to the division into two fragments. Quantitative this estimation can be carry out as follows:

With the nuclear distortion - without changing its volume - is converted into the ellipsoid with the axes

$a= R(1+\varepsilon)$
$b= R(1 - \varepsilon/2)$
$V= 4\pi R^3/3 = 4\pi ab^2/3$

With the deformation, the first member of a Weizsacke formula does not change, the second (surface energy) in the absolute value increase, but the third (Coulomb of energy) - decreases

$E_s = a_2 A^{2/3}(1+ 2\varepsilon^2 /5+....)$
$E_c = a_3 Z^2 A^{-1/3}(1 -\varepsilon^2 /5+....)$

Thus, the deformation changes the total energy of nucleus to value (further account the sign of the second and third terms in this formula

$$\Delta E= -\varepsilon^2/5(2a_2 A^{2/3} -a_3 \cdot Z^2 A^{-1/3}) \qquad \qquad \text{¤}$$

If the value of a change in the energy (¤) is positive, the nuclear binding energy will be increase, i.e., the deformation will be energetically advantageous and spontaneous fission is possible.

Dr. Mouaiyad M.S.Alabed

Consequently, fission barrier will disappear, when values (&) become more than zero, which begins with the values

$Z^2/A \geq 2a_2/a_3 \approx 48$

Follows to emphasize the approximate a nature got result as effects of the classical approach to quantum system - a nucleus

53. Estimate a change in the binding energy of a heavy nucleus in the fission of this nucleus into two identical nuclei of splinter. Consider the case

A= 240, Z= of 92

Solution:

In the division change the surface and Coulomb energies, their changes have the different signs:

$\Delta E_s = a_2 \cdot A^{2/3}(1-2^{1/3})$
$\approx -0,25.a_2.A^{2/3}$
$(17,2Mev) (240)^{2/3} (1-2^{1/3})$
$\approx -172,65Mev$

$\Delta E_c = a_3 \cdot Z^2(1- 2^{-2/3})/A^{1/3}$
$\approx 0,37.a_3.Z^2/A^{1/3}$
$\Delta Ec= (0,72Mev) (92)^2 (1-2^{-2/3}) /(240)^{1/3}$
$\approx 6094,08Mev (1-2^{-2/3}) / 6,214463$

Hence:

$\Delta Ec \approx 362,87Mev$

To the heavy nucleus with A= 240, Z= 92

$\Delta E_s \approx -172,65Mev, E_c$
$\approx 362,87Mev$

Sum" gain "in the energy with the fission of the heavy nucleus composes about 190Mev. This energy is expended, mainly, on the kinetic energies of nuclei -products

54. a) *Calculate the total binding energy and the binding energy per nucleon of $^3H_1$, $^4He_2$, and $^{31}Si_{14}$.*
   b) *Find the energy would be involved by removing the nucleons in Helium to infinity?*

Solution:

The total binding energy is:

$$E_{EB} = \{Zm_p + Nm_n - M_N\} \cdot 931,5Mev$$

For $^3H_1$:

$E_{EB} = \{1(1,007825) + 2(1,008665) - 3,016049\} \cdot 931,5Mev$
$E_{EB} = \{9,106.10\text{-}3\}.931,5Mev$
$E_{EB} = 8,482Mev$

For $^4He_2$:

$E_{EB} = \{2(1,007825) + 2(1,008665) - 4,002602\}.931,5Mev$
$\quad\quad = 0,030378 \cdot 931,5$
$E_{EB} = 28,297Mev.$
For $^{31}Si_{14}$:
$E_{EB} = \{14(1,007825) + 17(1,008665) - 30,975362\}. 931,5Mev$
$\quad\quad = 0,2815 .931,5$
$E_{EB} = 262,21Mev.$

The energy would be involved by removing the nucleons in Helium to infinity are:

28,297Mev which is enough to separate the nucleus of Helium to 2 protons and 2 neutrons.

55. Determine kinetic energies $E_{k\alpha}$ of α - particles, which are formed with α-disintegration $^{212}Bi_{83}$ to the excited states of nucleus $^{208}Tl_{81}$ with energies 0.49Mev and 0,61Mev. The binding energy E_{BE} of $^{212}Bi_{83}$ nucleus is 1654,32Mev, of $^{208}Tl_{81}$ nucleus is 1632,23Mev and of α - particle is 28,30Mev.

Solution:

Energy α - disintegration from the ground state of initial nucleus into the ground state of residual nucleus Q_0 is determined from the relationship:

Q_0= (M(A,Z) - M(A- 4,Z- 2) - M(α))c^2= E_{BE}(A- 4,Z- 2)+ E_{BE}(α) - E_{BE}(A,Z)

Where M(A.z)–the mass of the initial nucleus, M(A - 4, Z - 2) - the mass of the residual nucleus, M(α) – the mass of α - particle and E_{BE}(A,Z), $E_{BE.}$(A- 4,Z- 2), E_{BE}(α) with respect to their binding energy. In the general case, when disintegration occurs from the excited state of the initial nucleus into the excited state of the residual nucleus, the energy of α - disintegration is determined by the relation:

$Q= Q_0+ E_i - E_f$

Where E_i and E_f – the excitation energies of initial and final nucleus
Kinetic energy of α - particles taking into account the recoil energy of the residual nucleus:

$E_{k\alpha} = (Q_0+ E_i – E_f)$ {M(A-4,Z-2)/M(A-4,Z-2)+M(α)}
    $\approx (Q_0+E_i-E_f)${(A-4)/A}

Upon decay to the first excited state (0.49Mev) of nucleus ^{208}Tl

$E_{k\alpha}$= (1632,23+28,30–1654,32–0,49)Mev.208amu/212amu
    =5,61Mev

In the disintegration to the second excited state (0,61Mev) the energy of α - particles will be?

$E_{k\alpha}$= (1632,23+28,30 – 1654,32 – 0,61) Mev. 208 amu/212 amu
=5,49Mev

56. Determine the recoil energy of nucleus ^7Li, which is formed with e capture in the nucleus of ^7Be 4

Solution:

Are given the nuclear binding energy:

$E_{BE}(^7Be)$= 37.6 Mev, $E_{BE}(^7Li)$= 39.3 Mev

Process:

$^7Be + e^-$ $\longrightarrow$ $^7Li + \nu_e$

The energy of e-capture is

$Q_e = E_{BE}(A, Z-1) - E_{BE}(A, Z) - (m_n - m_p)c^2 + m_e c^2 = E_{BE}(A, Z-1) - E_{BE}(A, Z) - 0.78$ Mev

Where $E_{BE}(A, Z)$ and $E_{BE}(A, Z-1)$ are the binding energies of initial and residual nuclei and m_n, m_p and m_e are the masses of the neutron, proton and the electron:

$Q_e = E_{BE}(^7Li) - E_{BE}(^7Be) - 0.78$Mev
= (39,3-37,6-0,78)Mev $\simeq$ 0,9Mev

From the laws of conservation of the energy and the momentum follows where:

$T_{Li} + T_\nu = Q_e$
$| p_{Li}| = | p_\nu|,$

Where: T_{Li} , T_v are the kinetic recoil energies of nucleus and neutrino. Neutrino - relativistic particle, and nucleus-nonrelativistic:

$p_{Li} = \{\sqrt{2m_{Li}} c^2 T_{Li}\}/c^2$ $p_v = (T_v/c)$

In the final step, we have:

$T_{Li} = T^2_v / 2m_{Li} c^2$
$\approx Q^2_e / (2A_{Li}.931)$
$= (0,9Mev)^2/(2. 7. 931Mev)$
$\approx 6.10^{-5}Mev = 60eV$

57. Determine the kinetic energy of residual nucleus with β^- nuclear decomposition:

$^{64}Cu_{29}$ $(^{64}Cu_{29} \rightarrow {}^{64}Zn_{30} + e + \acute{v}_e)$ when

1) Energy of antineutrino $T_{\acute{v}} = 0$,
2) Energy of electron $T_e = 0$.

The nuclear binding energy of $^{64}Cu_{29}$ is 559,32Mev and for $^{64}Zn_{30}$ is 559,12Mev.

Solution:

Energy of β^- disintegration:

$Q_{\beta^-} = E_{BE}(A,Z+1) - E_{BE}(A, Z) + (m_n - m_p)c^2 - m_e c^2$
$E_{BE}(A,Z+1) - E_{BE}(A,Z) + 0.78$ Mev
$= 0.58Mev$

Where $E_{BE}(A, Z)$ and $E_{BE}(A, Z+ 1)$ – the initial binding energy and residual nuclei; m_n, m_p and m_e - mass of the neutron, the proton and the electron. Recoil energy of nucleus with β^- -disintegration will be 1)$T_{\acute{v}} = 0$. Let us write down the laws of conservation of energy and momentum

$$Q_{\beta-} = E_{kZn} + E_{kv\backslash} + E_{ke} = E_{kZn} + E_{ke}$$

$$|p_{Zn}| = |p_e|$$

For the pulses, taking into account that P_{Zn} - nonrelativistic pulse, p_e - relativistic pulse:

$$p_{Zn} = 1/c\{\sqrt{(2m_{Zn}c^2 E_{kZn})}\} \ , \ p_e = (1/c)\{\sqrt{(E_k^2{}_e + 2E_{ke}m_ec^2)}$$

It is possible to write down where m_{Zn} - nuclear mass ^{64}Zn. From the laws of conservation we have:

$$2m_{Zn}c^2 E_{kZn} = E_k^2{}_e + 2E_{ke}m_ec^2 =$$
$$= E_{ke}(E_{ke} + 2m_ec^2) = (Q_{\beta-} - E_{kZn})(Q_{\beta-} - E_{kZn} + 2me))$$

Since $m_e \ll m_{Zn}$, то $E_{kZn} \ll Q_\beta$

$$E_{kZn} \approx Q_{\beta-}(Q_{\beta-} + 2m_e)/2m_{Zn}c^2$$
$$\approx Q_{\beta-}(Q_{\beta-} + 2m_e)/(2A_{Zn}.931) =$$
$$= \{0,58.(0,58 + 2.0,51)Mev^2\}/(2.64.931Mev)$$
$$\approx 7,8eV$$

3) $E_ke = 0$. Analogously as in the first case:

$$Q_{\beta-} = E_{kZn} + E_{kv}$$

$$|p_{Zn}| = |p_v|$$

The pulse of antineutrino ultra-relativistic

$$P_v = E_{kv\backslash}/c$$

Finely:
We will obtain

$$E_{kZn} \approx Q^2{}_{\beta-}/2m_{Zn}c^2$$
$$\approx Q^2{}_{\beta-}/(2.64m_{Zn}c^2)$$
$$= 0,58^2/2.64.931 \approx 2,8eV$$

58. Find the separation energy of an alpha particle from $^{12}C_6$

Solution:

$B_a =$ M(A-4,Z-2)+ M(^{4}He$_2$) - M(A,Z)
$\quad$ =Δ (A-4,Z-2)+Δ (^{4}He$_2$) -Δ (A,Z)
$B_a =$ (4.941+ 2.424)Mev= 7.365Mev

Let us compare the results, obtained for the specific nuclear binding energy of $^{12}C_6$ and energies of the separation of neutron, proton and alpha particle from this nucleus. Energy of the separation of one nucleon from this nucleus proved to be more than doubly higher than specific binding energy! Energy of the simultaneous separation of cluster of 4 nucleons and α- particle - less than the specific binding energy - i.e. the average of the separation energy of one nucleon. These facts and analogous results for a number of other nuclei became the basis of the theoretical model of nuclear shells.

59. The masses of a neutron and of a proton in the energy units are respectively m_n= 939,6 Mev and m_p= 938,3 Mev. Determine nuclear mass $^2H_1$ in the energy units, if the binding energy of deuteron is E_{BE}= 2,2 Mev

Solution:

Mass of the nucleus M(A,Z)= Zm_p+ (A–Z)m_n – E_{BE}(A,Z)
Where Z and A - respectively charge and nuclear mass

Then:

For the deuteron

M = 1.938,3Mev+1. 939,6Mev–2,2Mev
$\quad$ =1875,7Mev

60. Determine the specific nuclear binding energy of $^{16}O_8$ nucleus When the mass of the neutral atom of $^{16}O_8$ M is 15,9949amu

Solution:

When the specific nuclear binding energy of the nucleus is:

$\varepsilon (A,Z) = E_{BE} / A$

Where $E_{BE}(A,Z)$ - the nuclear binding energy, A - mass number

The total energy of the connection of the nucleus:

$E_{BE} = [Zm_p + (A-Z)m_n - M_N]c^2$
$\quad\quad = \{Zm_p + (A-Z)m_n - M_{at} - Zm_e\}c^2$

The using of the energy units for the masses is 1amu= 931,5Mev, we obtain for ^{16}O nucleus

$\varepsilon = [8.(m_p.931,5) + 8(m_n.931,5) - [M_{16O}. 931,5] - 8.m_e] / 16$
$\varepsilon = [8.(1,007825.931,5) + 8(1,008665.931,5) - [15,994915.$
$\quad\quad 931,5] - 8.0,511] / 16$
$\varepsilon = [8. 938,27\text{Mev} + (16-8). 939,57\text{Mev} - 15,9949. 931,49\text{Mev} -$
$\quad\quad 8.0,511\text{Mev}] / 16$
$\quad = 131,7079/16$
$\varepsilon = 8,232$ Mev/nucleon

61. Calculate the energies of the neutrons branches in even-even isotopes $^{38}Ca_{20}$, $^{40}Ca_{20}$, $^{48}Ca_{20}$ with the aid of Weizsacker's formula

Solution:

The energy of the neutron branches in the nucleus (A,Z) is:

$\varepsilon_n(A,Z) = [m_n + M(A-1,Z) - M(A,Z)]c^2$

The nucleus mass is:

$Mc^2 = [Zm_p + (A-Z)m_n]c^2 - E_{BE}$

The energies of the neutron branches are:

$\varepsilon_n(A,Z) = [m_n + Zm_p + (A-1-Z)m_n]c^2 - E_{BE}(A-1,Z) - [Zm_p + (A-Z)m_n]c^2 + E_{BE}(A,Z) - E_{BE}(A-1,Z)$

The energy of the binding of atomic nuclei is described with the aid of the formula of Weizsacker:

$E_{BE}(A,Z) = a_1 A - a_2 A^{2/3} - a_3 \{Z(Z-1)/A^{1/3}\} - a_4 \{A/2 - Z\}^2 /A\} + a_5(1/A^{3/4})$

Where:

$a_1 = 15.78$ Mev
$a_2 = 17.8$ Mev
$a_3 = 0.71$ Mev
$a_4 = 94.8$ Mev
$a_5 = 0$

For the nuclei with odd A, $a_5 = +34$ Mev for the even even-mass nuclei and

$a_5 = -34$ Mev for the odd -odd nucleus. Then for the nuclei (A,Z) binding energy will be:

^{38}Ca:

$E_{BE}(38,20) = 15,78Mev.38 - 17,8Mev. 38^{2/3} - 0,71Mev.[20(20-1)/38^{1/3}] - 94,8Mev[(38/2 -20)^2 /38] + 34Mev / 38^{3/4} = 317,9Mev$

^{40}Ca:

$E_{BE}(40,20) = 15,78Mev.40 - 17,8Mev. 40^{2/3} - 0,71Mev.[20(20-1)/40^{1/3}] + 34Mev / 40^{3/4} = 346,3Mev$

^{48}Ca:

E_{BE} (48,20)= 15,78Mev.48 – 17,8Mev. $48^{2/3}$ – 0,71Mev.[20(20-1) / $48^{1/3}$] – 94,8Mev[(48/2 -20)^2 /48]+ 34Mev / $48^{3/4}$ = 418,4Mev
For the nucleus (A-1,Z) the binding energy will be:

^{37}Ca:

E_{BE} (37,20)= 15,78Mev. 37 – 17,8Mev. $37^{2/3}$ – 0,71Mev. [20(20-1)/$37^{1/3}$] – 94,8Mev. [(37/2-20)^2/37] = 299,5Mev

^{39}Ca:

E_{BE}(39,20)= 15,78Mev. 39 – 17,8Mev. $39^{2/3}$ – 0,71Mev.[20(20-1)/$39^{1/3}$] – 94,8Mev. [(39/2-20)^2/39] = 330,6Mev

^{47}Ca:

E_{BE} (47,20)= 15,78Mev. 47 – 17,8Mev. $47^{2/3}$ – 0,71Mev.[20(20-1)/$47^{1/3}$] – 94,8Mev. [(47/2-20)^2/47] = 410,3Mev

The energy of the separation of a neutron is:

^{38}Ca ε_n(38,20)= 317.9 Mev - 299.5 Mev= 18.4 Mev
^{40}Ca ε_n(40,20)= 346.3 Mev - 330.6 Mev= 15.7 Mev
^{48}Ca ε_n(48,20)= 418.4 Mev - 410.3 Mev= 8.1 Mev

62. Calculate radii of the specular nuclei $^{23}Na_{11}$, $^{23}Mg_{12}$. When the binding energy of these nucleuses are:

$E_{BE}(^{23}Na_{11})$= 186,56 Mev, $E_{BE}(^{23}Mg_{12})$= 181,72 Mev

Solution:

The Coulomb's energy of the evenly charged sphere of radius R is defined by relationship:

Where $a_1 = 15.78$ Mev

$a_2 = 17.8$ Mev

$a_3 = 0.71$ Mev

$a_4 = 94.8$ Mev

$a_5 = 0$ for the nuclei with odd A

$a_5 = +34$ Mev for the even even-mass nuclei and

$a_5 = -34$ Mev for the odd nuclei

The last term $a_5/A^{3/4}$ as a result of its smallness to consider not there. With the division of initial nucleus(A_{in} , Z_{in}) into two identical splinters(A_{sp}, Z_{sp}) their masses numbers and charges have the following relationships: $A_{sp} = A_{in}/2$ and $Z_{sp} = Z_{in}/2$

The fission energy of nucleus will depend only on the second and third members of the formula of Weizsacker - the surface and Coulomb energy:

$$Q_f = W^{su}_{in} + W^{col}_{in} - 2W^{su}_{sp} - 2W^{col}_{sp}$$

The surface energy of splinters is:

$$2W^{su}_{sp} = 2a_2 A^{2/3}_{sp} = 2a_2(A_{in}/2)^{2/3}$$
$$2(1/2)^{2/3}a_2 A_{in}^{2/3} = 2^{1/3}W^{su}_{in} = 1,26\ W^{su}_{in}$$

Coulomb energy of splinters is:
$$2W^{col}_{sp} = (2a3)Z^2_{sp}/A^{1/3}_{sp}$$
$$= (2a_3)(Z_{in}/2)^2/(A_{in}/2)^{1/3}$$
$$= 2.2^{1/3}.(1/2)^2 a_3(Z^2_{in}/A^{1/3}_{in})$$
$$= 2^{-2/3}W^{col}_{in} = 0,63E^{col}_{in}$$

The energy of nuclear fission Q_f is separated as a result of changing the Coulomb and surface energy of initial nucleus and splinters:

$$Q_f = 0,37W^{col}_{in} - 0,26W^{su}_{in} = 0,37a_3(Z^2_{in}/A^{1/3}_{in}) - 0,26a_2 A^{2/3}_{in}$$
$$= \{0,37.0,71Mev.92^2] / 238^{1/3} - 0,26.\ 17,8Mev.238^{2/3}$$
$$\approx 360Mev - 180Mev$$
$$Q_f = 180Mev$$

65. Does appear reaction $^6Li_3(d,\alpha)^4He_2$ of endothermic or exothermic, when we are given the specific nuclear binding energy inMev:

$$\varepsilon(d)= 1,11; \; \varepsilon(\alpha)= 7,08; \; \varepsilon(^6Li_3)= 5,33$$

Solution:

By using the relations of specific nuclear binding energy, we can find the energy of the reaction as:

$Q= 2E_\alpha(^4He) - E_\alpha(^6Li) - E_\alpha(^2H)$
$=2A_\alpha\varepsilon_\alpha - A_{Li}\varepsilon_{Li} - A_d\varepsilon_d=$
$=2(4)(7,08)-(6)(5,33)-(2)(1,11)$
$=56,64-31,98-2,22$
$Q= 22.44$ Mev

Hence:
The reaction is endothermic

66. What are the total binding energy and the average of the binding energy per particle of $^{235}U_{92}$ nucleus?

Solution:

The atomic rest mass of this nucleus is: 235,043924amu

When $m_p= 1,007825$amu

And $m_n= 1,008665$amu

The total mass of all protons in this nucleus is:

$Zm_p= 92. \, 1,007825= 92,7199$amu

And $Nm_n= 143. \, 1,008665= 144,23909$amu

Thus the total binding energy of $^{235}U_{92}$ is 235,043924amu

(92,7199+ 144,23909) - 235,043924= 1,91507amu

The total binding energy is:

1,91507. 931,5= 1783,8877Mev.

And the average binding energy per particle is:

1783,8877 / 235 = 7,59Mev per particle.

67. What is the binding energy of the isotope $^{210}Pb_{82}$? What is the binding energy per nucleon?

Solution:

The total binding energy inMev is:

$[(Zm_p + Nm_n) - M_N]$. 931,5

When M_N is the mass of $^{210}pb_{82}$ = 209,984163amu And m_p is 1,007825amu

And m_n = 1,008665amu

Hence the total binding energy is:

$[(82. 1,007825)+ (128. 1,008665) - 209,984163]$. 931,5
 $= [82,64165+ 129,10912 - 209,984163]$. 931,5
 $= 1,76661. 931,5 = 1645,5972$Mev.

And the binding energy per nucleon is:
 1645,5972 / 210 = 7,836Mev per nucleon.

68. Calculate the binding energy of the last neutron in a $^{12}C_6$ nucleus. (Hint: compare the mass of $^{12}C_6$ with that of $^{11}C_6 + {}^1n_0$ use appendix).

Solution:

The protons number of this nucleus is 6 and the neutrons number of this nucleus is 6 too, hence:

$(6m_p + 6m_n) = 6 (1,007825 + 1,008665)$
$= 12,09894$amu.

And $M_{12B6} = 12$amu

$\Delta m = 12,09894 - 12 = 0,09894$amu

Hence the binding energy of $^{12}B_6$ is:

$0,09894 . 931,5 = 92,16261$Mev.

Binding energy of $^1n_0 = 92,16261 / 6 = 15,360435$Mev.

$M_{11C6} = 11,011433$amu
$6m_p + 5m_n = 6 . 1,007825 + 5 . 1,008665$
$= 6,04695 + 5,043325 = 11,090275$amu
Δm of $M_{11C6} = 11,090275 - 11,011433 = 0,078842$amu

And the binding energy of $^{11}C_6 = 0,078842 . 931,5 = 73,441323$Mev.

The sum of $^{11}C_6$ nuclied and a neutron is:

$15,360435 + 73,441323 = 88,801758$Mev.

The difference between them is:

$92,16261 - 88,801758 = 3,360852$Mev.

69. Calculate the total binding energy and the binding energy per nucleon for $^6Li_3$ Use Appendix.

Solution:

The protons number of this nucleus is 3 and the neutrons number of this nucleus is 3 too, hence:

$(3m_p + 3m_n) = 3(1,007825 + 1,008665)$
$= 6,04947$ amu

And $M_{6Li3} = 6,015121$ amu

$\Delta m = 6,04947 - 6,015121 = 0,034349$ amu

Hence the binding energy of $^6Li_3$ is:

$\Delta m \cdot 931,5 = 0,034349 \cdot 931,5 = 31,996093$ Mev.

The binding energy per nucleon is:

$31,996093 / 6 = 5,3327$ Mev / nucleon.

70. What is the binding energy of the isotope $^{210}Pb_{82}$? What is the binding energy per nucleon for this isotope?

Solution:

The total binding energy in Mev is:

$[(Zm_p + Nm_n) - M_N] \cdot 931,5$

When M_N is the mass of $^{210}Pb_{82} = 209,984163$ amu
And m_p is $1,007825$ amu And $m_n = 1,008665$ amu

Hence the total binding energy is:

[(82 . 1,007825)+ (128 . 1,008665) - 209,984163] . 931,5
= [82,64165+ 129,10912 - 209,984163] . 931,5
= 1,76661 . 931,5 = 1645,5972Mev.

And the binding energy per nucleon is:

1645,5972 / 210 = 7,836Mev per nucleon.

71. What are the total binding energy and average binding energy per particle of nucleus $^{235}U_{92}$?

Solution:

The atomic rest mass of this nucleus is 235,043924amu

When m_p = 1,007825amu
And m_n = 1,008665amu
The total mass of all protons in this nucleus is:

Zm_p = 92 . 1,007825 = 92,7199amu

And Nm_n = 143 . 1,008665 = 144,23909amu
Thus the total binding energy of $^{235}U_{92}$ is: 235,043924amu
Hence (92,7199+ 144,23909) − 235,043924 = 1,91507amu
Hence the total binding energy is: 1,91507 . 931,5 = 1783,8877Mev.

And the average binding energy per particle is:

1783,8877 / 235 = 7,59Mev per nucleon.

72. A nitrogen $^{14}N_7$ nucleus absorbs a deuterium $^2H_1$ nucleus during a nuclear reaction. What is the name, atomic number and

nucleon number of the compound nucleus and find the type of this reaction.

Solution:

The reaction is:

$$^{14}N_7 + {}^2H_1 \rightarrow {}^{16}O_8 + \gamma$$

We find the γ's energy, that is the difference between the energy of reaction materials and the energy of daughters nuclide:

For $^{14}N_7$:
$7(m_p) + 7(m_n) =$
$$= 7(1,007825) + 7(1,008665)$$
$$= 7,054775 + 7,060655$$
$$= 14,11543 \text{amu}$$

And the atomic mass of $^{14}N_7$ is 14,003074amu

Thus:

$\Delta m = 14,11543 - 14,003074$
$$= 0,112356 \text{amu}$$

And the energy of $^{14}N_7$ reaction is:

$E_{BE} = 0,112356 . 931,5$
$$= 104,65961 \text{Mev}$$

For $^2H_1$:
$1m_p + 1m_n = 1,007825 + 1,008665$
$$= 2,01649 \text{amu}$$

And the atomic mass of $^2H_1$ is 2,014102amu

Hence:

$\Delta m = 2,01649 - 2,014102$
$= 2,388. \ 10^{-3}$amu

And the binding energy of $^2H_1$ is $2,388.10^{-3}. \ 931,5 = 2,224422$Mev.

The energy of the reaction materials is:

$104,65961 + 2,224422 = 106,88403$Mev

The energy of results nuclide is:

$8m_p + 8m_n = 8(1,007825) + 8(1,008665)$
$= 8,0626 + 8,06932$
$= 16,13192$amu

And the atomic mass of $^{16}O_8$ is $15,994915$amu

Hence:

$\Delta m_0 = 16,13192 - 15,994915$
$= 0,137005$amu

And:

$E = 0,137005. \ 931,5$
$= 127,62015$Mev.

Hence:

$Q = 127,62015 - 106,88403$
$= 20,73612$Mev.
Thus the energy of γ is $20,736$Mev.

And the reaction becomes:

$^{14}N_7 + {}^2H_1 \rightarrow {}^{16}O_8 + \gamma \ (20,736\text{Mev.})$

And we can say that:

$\Delta m = 14{,}003074 + 2{,}014102 - 15{,}994915$
$= 0{,}0222609 amu$

And the energy is:

$0{,}0222609 . 931{,}5 = 20{,}73612 Mev.$

73. Two isotopes of the same element have the same binding energy. One isotope contains two more neutrons than the other. What is the difference between the atomic masses (in atomic mass units) of these isotopes?

Solution:

We write the symbols of these isotopes as:

$^{A1}_{Z1}X_n \ ^{A2}_{Z2}Y_{n+2}$ when $Z_1 = Z_2$
$^{A1}_{Z}X_n \ ^{A2}_{Z}Y_{n+2}$
M_X atomic mass for X => $M_X = Zm_p + Nm_n$
M_Y atomic mass for Y => $M_Y = Zm_p + (N+2)(m_n)$
$(BE)_X = M_X - (Zm_p + Nm_n)$
$(BE)_Y = M_Y - [Zm_p + (N+2)m_n]$
$(BE)_X = (BE)_Y$

Then:

$M_X - (Zm_p + Nm_n) = M_Y - [Zm_p + (N+2)m_n]$
$-M_Y + M_X = Zm_p + Nm_n - Zm_p - Nm_n - 2m_n$

Hence:

$M_Y - M_X = 2m_n$

Hence:

$\Delta M = 2(1,008665)$amu

Hence:

$\Delta M = 2,01733$amu

74. Calculate the binding energy of $^{40}Ca_{20}$ nuclide with using the following scheme:

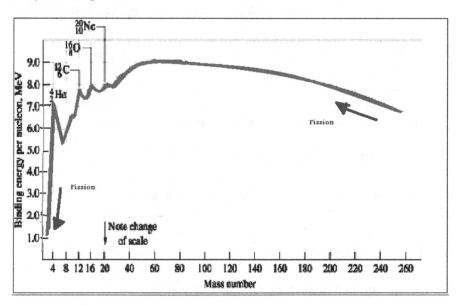

Solution:

The number of the protons in this nuclide is:

$20m_p = 20(1,015650\text{amu}) = 20,313$amu

And the number of the neutrons in this nuclide is:

$20m_n = 20.(1,01733 \text{ u.}) = 20,3466$amu

Hence the number of nucleons in this nucleus is:

$20m_p + 20m_n = 20{,}313amu + 20{,}3466amu =$
$$= 40{,}6596amu$$

$\Delta m = 40{,}6596 - 39{,}962591amu =$
$$= 0{,}697009amu$$
$E_{BE} = \Delta m. \ 931{,}5Mev/amu = 649{,}26388Mev.$

75. Calculate the: a) Total binding energy of $^{238}U_{92}$ nucleus.
b) Total binding energy of $^{107}Ag_{47}$ nucleus.

Solution:

a) $^{238}U_{92} =>$

The number of the protons in this nuclide is:

$92m_p = 92(1{,}015650amu) = 93{,}4398amu$

And the number of the neutrons in this nuclide is:

$146m_n = 146.(1{,}01733amu) =$
$$= 148{,}53018amu$$

Hence the number of nucleons in this nucleus is:

$92m_p + 146m_n = 93{,}4398amu + 148{,}53018amu =$
$$= 241{,}96998amu$$

$\Delta m = 241{,}96998amu - 238{,}050784amu = 3{,}9192amu$
$E_{BE} = \Delta m. \ 931{,}5Mev/amu =$
$$= 3{,}9192 \ u.. \ 931{,}5Mev/amu =$$
$$= 3650{,}7348Mev.$$

b) $^{107}Ag_{47} =>$

The number of the protons in this nuclide is:

$47m_p = 47(1,015650amu) = 47,73555amu$

And the number of the neutrons in this nuclide is:

$60m_n = 60.(1,01733amu) = 61,0398amu$

Hence the number of nucleons in this nucleus is:

$47m_p + 60m_n = 47,73555amu + 61,0398amu =$
$$= 108,77535amu$$

$\Delta m = 108,77535amu - 106,905091amu$
$$= 1,87026amu$$
$E_{BE} = \Delta m. \; 931,5Mev/amu$
$$= 1,87026amu . \; 931,5Mev/amu$$
$$= 1742,1471Mev.$$

76. a) *Determine that the nucleus $^8Be_4$ is instable to decay to two alpha particles,*
   b) *Is the isotope $^{12}C_6$ can decay to three alpha particles? If it is yes, why? And if it is no, why?*

Solution:

The number of the protons in the $^8Be_4$ nuclide is:

$4m_p = 4(1,007825amu) = 4,0313amu$

And the number of the neutrons in this nuclide is:

$4m_n = 4.(1,008665amu) = 4,03466amu$

Hence the number of nucleons in this nucleus is:

$4m_p + 4m_n = 4,0313$amu$+ 4,03466$amu
$= 8,06596$amu

And to alpha particle is:

$2m_p + 2m_n = 4,03298$amu

And the mass of $2\alpha = 8,06596$amu

The binding energy of the nucleus $^8Be_4$ is:

$(8,06596 - 8,005308).\ 931,5=$
$= 0,060652$amu$.\ 931,5$
$= 56,4973387$Mev

The binding energy of 2α is:

$E_{BE(2\alpha)} = 8,06596amu-2(4,002602).\ 931,5$
$= 56,594214$Mev.

Because the binding energy of the $^8Be_4$ nucleus is < the binding energy of two alpha particles, hence the first cannot decay to two alpha particles.

b) because of the binding energy of:

$^{12}C_6$ per neucleon$= 184,32522$Mev$/12$
$= 15,360435$Mev/nucleon

The mass of three alpha particles is: $12,007806$amu

The Δm of alpha particle is:

$(4,03298 - 4,002602) = 0,030378$amu

And the binding energy of three alpha particles is:

3(0,030378 u.). 931,5Mev/amu=
$$= 0,091134. 931,5$$
$$= 84,891321 \text{Mev}.$$

Then: 184,32522Mev > 84,891321Mev
Hence this nucleus can decay to three alpha particles.

77. Is the following reaction occurs or not? Why?

$$^{11}C_6 \rightarrow {}^{10}B_5 + p$$

Solution:

The mass of the $^{11}C_6$ nucleus is 11,011433amu

The mass of the $^{10}B_5$ nucleus is 10,012936amu and of the proton is 1,007825amu

10,012936amu+ 1,007825amu= 11,020761amu

Therefore the reaction energy Q is:

$Q = 11,011433$amu - 11,020761amu
$= -9,328.10^{-3}. 931,5$
$Q = -8,689$Mev.

The negative sign indicate that the reaction cannot be happen because the energy is not be enough to make it.

78. Find the total binding energy and the average binding energy per nucleon for the $^{90}Sr_{38}$ nucleus.

Solution:

The total atomic mass for protons in the nucleus is:

Zm_p = 38. 1,007825amu = 38,29735amu

And for neutrons in the nucleus is:

Nm_n = 52. 1,008665amu = 52,45058amu

The atomic mass of the nuclide in the nucleus is:

$Zm_p + Nm_n$ = 38,29735amu+ 52,45058amu= 90,74793amu

The atomic mass of the nucleus $^{90}Sr_{38}$ is:

M_N = 89,907737amu
Δm = 90,747934amu - 89,907737amu = 0,840197amu

Therefore the total binding energy is:
E_{BE} = 0,840197 . 931,5 = 782,6435Mev.

The average of binding energy per nucleon is:

$E_{BE.p}$ = 782,6435 / 90 = 8,6960388Mev per nucleon.
$E_{BE.p} \approx$ 8,7Mev.

79. Find the binding energy of the isotope $^{152}Sm_{62}$ and find the binding energy of this isotope per nucleon?

Solution:

m_p= 1,007825amu , m_n= 1,008665amu , M_N= 151,919728amu
Zm_p= 62. 1,007825amu= 62,48515amu
Nm_n= 90. 1,008665amu= 90,77985amu

Therefore:

The binding energy is:

$E_{BE} = \{(Zm_p + Nm_n) - M_N\}. 931,5\text{Mev/amu}$
$= \{(62,48515 + 90,77985) - 151,919728\}. 931,5$
$= 1,34528. 931,5 = 1253,1283\text{Mev.}$

The binding energy of the isotope per nucleon:

$E_{BE} = 1253,1283 / 152 = 8,2442651\text{Mev / nucleon}$

80. Calculate the binding energy with the nucleon units to the nucleus of $^{12}C_6$.

Solution:

The number of the protons in this nuclide is:

$6m_p = 6(1,015650\text{amu}) = 6,0939\text{amu}$

And the number of the neutrons in this nuclide is:

$6m_n = 6.(1,01733\text{amu}) = 6,10398\text{amu}$

Hence the number of nucleons in this nucleus is:

$6m_p + 6m_n = 6,0939\text{amu} + 6,10398\text{amu}$
$= 12,19788\text{amu}$

$\Delta m = 12,19788\text{amu} - 12,00000\text{amu}$
$= 0,19788\text{amu}$
$E_{BE} = \Delta m. 931,5\text{Mev/amu}$
$= 0,19788\text{amu}. 931,5$
$= 184,32522\text{Mev.}$

The total binding energy per nucleon is:

$$E_{BE}/A = 184{,}32522\text{Mev} / 12$$
$$= 15{,}360435\text{Mev} / \text{nucleon}.$$

81. Calculate the binding energy of the last neutron in the nucleus of $^{12}C_6$ (compare the mass of $^{12}C_6$ with the produced mass to $^{11}C_6 + {}^1n_0$).

Solution:

The number of the protons in the $^{11}C_6$ nuclide is:

$$6m_p = 6(1{,}015650\text{amu}) = 6{,}0939\text{amu}$$

And the number of the neutrons in this nuclide is:

$$6m_n = 6.(1{,}01733\text{amu}) = 6{,}10398\text{amu}$$

Hence the number of nucleons in this nucleus is:

$$6m_p + 6m_n = 6{,}0939\text{amu} + 6{,}10398\text{amu}$$
$$= 12{,}19788\text{amu}$$

And the number of the protons in this $^{11}C_6$ nuclide is:

$$6m_p = 6(1{,}015650\text{amu}) = 6{,}0939\text{amu}$$

And the number of the neutrons in this nuclide is:

$$5m_n = 5.(1{,}01733\text{amu}) = 5{,}08665\text{amu}$$

Hence the number of nucleons in this nucleus is:

$$6m_p + 5m_n = 6{,}0939\text{amu} + 5{,}08665\text{amu}$$
$$= 11{,}18055\text{amu}$$

The atomic mass of $^{11}C_6$ = 11,009305amu

Hence $m(^{11}C_6) + m_n$ = 11,009305amu + 1,01733amu
$$= 12,026635amu$$

And the binding energy of the nucleus of $^{16}C_6$ is:

$(6m_p + 5m_n) - M(^{11}C_6)$ = 11,18055amu - 11,009305amu
$$= 0,171245amu$$

The binding energy of a neutron is:

[0,171245 / 16]. 931,5 = 9,9696693amuMev / nucleon.

And the mass of the neutron m_n = 1,01733amu

Hence:

The binding energy of the last neutron in the nucleus of $^{12}C_6$ is:

$E_{BE.n}$ = 9,9696693 / 1,01733
$$= 9,7998381Mev.$$

82. The isotope of $^{22}Na_{11}$ in a tissue an approximate 0,01grams per kilogram total tissue ,and the isotope of $^{40}K_{19}$ in this tissue at an approximate 0,008grams per kilogram total tissue. Calculate the radioactivity of two isotopes in this tissue, when the mass of this tissue is 2 kg.

Solution:

The mass of the isotope $^{22}Na_{11}$ in 2kg of the tissue is:

0,01. 2kg = 0,02 kg

And the mass of the isotope $^{40}K_{19}$ in 2kg of the tissue is:

0,008. 2kg = 0,016 kg

The number of atoms of the isotope of $^{22}Na_{11}$ in this tissue is:

$N = (6,022.10^{23} / 22). 0,02=$
$\quad = 5,4745.10^{20}$ atoms.

The radioactivity of this isotope is:

$A = \lambda N$
$\quad = (0,693 / 2,6088.365,25.24.3600). 5,4745.10^{20}$
$\quad = 8,4176.10^{-9}. 5,4745.10^{20}$
$A = 4,608.10^{12}$ Bq

And the number of atoms of the isotope of $^{40}K_{19}$ in this tissue is:

$N = (6,022.10^{23} / 40). 0,016$
$\quad = 2,4088.10^{20}$ atoms.

The radioactivity of this isotope is:

$A = \lambda N$
$\quad = (0,693 / 1,227.10^9.365,25.24.3600). 2,4088.10^{20}$
$\quad = 1,7897.10^{-17}. 2,4088.10^{20}$
$A = 4311,0734$Bq

83. What is the mass of 1 μCi of $^{98}Tc_{43}$ ($T_{1/2}= 4,2.10^6yr$)?

Solution:

The decay constant of this isotope is:

$\lambda = 0,693 / T_{1/2}$
$\quad = 0,693 / \{4,2.10^6. 365,25.24.3600\}$
$\quad = 5,228.10^{-15}$ sec^{-1}

The activity of this isotope is:

$A = \lambda N$

$1\mu Ci \cdot 10^{-6} = 5,228 \cdot 10^{-15} \ sec^{-1} \cdot N$

Hence:

$N = 10^{-6} / 5,228 \cdot 10^{-15} = 1,9127 \cdot 10^{8}$ nuclei.

$N / mass = N_A / A$

$1,9127 \cdot 10^{8} / m = 6,025 \cdot 10^{23} / 98$

$m = 3,11 \cdot 10^{-14}$ gram.

84. A 0,018 μCi of a sample of $^{32}P_{15}$ is injected into an animal for tracer studies. If a Geiger counter intercepts 20 per cent of the emitted β particles and is 90 percent efficient in counting them, what will be the counting rate?

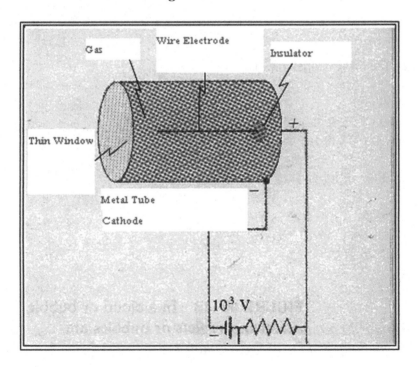

Solution:

$A = \lambda N$
$0,018. 10^{-6}.3,7.10^{10} = (0,693 / 14,262.24.3600). N$
$N= 666 / 5,6239.10^{-7}$
$N= 1,1842.10^{9}$ nuclei.
$(20/100). 1,1842.10^{9}= 2,368.10^{8}$ beta particles which emitted.
$(90/100). 1,1842.10^{9}= 1,06578.10^{9}$ .... in counting

Therefore:

Counting rate= $(2,368.10^{8}+ 1,06578.10^{9})/2 = 6,513.10^{8}$ beta particles.

85. A sample of the nucleus of $^{14}C_6$ which has 1.10^{21} nuclei and its half-life is 5730 yr. Calculate its radioactivity.

Solution:

When the decay constant λ is:

$\lambda = 0,693 / T_{1/2}$
$= 0,693 / 5730.365,25.24.3600$
$= 0,693 / 1,808 \ 10^{11}= 3,8324.10^{-12}$ sec^{-1}

The radioactivity is:

$A = \Delta N / \Delta t$
$= \lambda N = (3,8324.10^{-12})(1.10^{21})$
$= 3,8324.10^{9}$ dis/sec.

86. Obtain the relationship between the half-life period, disintegration probability and the average of half-life

Solution:

The period of half-life is called the time, for which the number of particles (or nuclei) will decrease double:

When: $N_{T1/2} = N_0 \, e^{-\lambda T1/2}$

And: $N_{T1/2} = N_0 / 2$ $\qquad\qquad\qquad$ (@@@)

Hence:

$N_0 / 2 = N_0 \, e^{-\lambda T1/2}$
$e^{-\lambda T1/2} = 1/2$
$\ln 2 = \lambda T_{1/2}$

When:

$\tau = 1 / \lambda$

Therefore:

$T_{1/2} = \ln 2 / \lambda = \tau \ln 2$

87. Find the kinetic energy of α- particle and recoil nuclei in the disintegration of radium

Solution:

The form of this reaction is:

$$^{226}Ra_{88} \quad \rightarrow \quad ^{222}Rn_{86} + {}^{4}He_{2}$$

A difference in the masses of radium and products of its:

$\Delta M = M(226, 88) - M(222, 86) - M(4, 2)$
$\quad = \Delta(226,88) + 226 - \Delta(222,86) - 222 - \Delta(4,2) - 4$

$=\Delta(226,88)-\Delta(222,86)-\Delta(4,2)$
$=(23.662-16.367-2.424)Mev=4.87Mev$

Here were used the tables of the difference of the masses of neutral atoms

The kinetic energy is:

$E_{k\alpha}=(\Delta M. M_{Rn})/(M_{\alpha}+M_{Rn})$
$(4,87Mev.222)/226=4,79Mev$
$E_{kRn}=(4,87Mev.4)/226=0,09Mev$

88. Determine the activity of the preparation of gold – 198($^{198}Au_{79}$), induced during irradiation of the model of gold - 197($^{197}Au_{79}$) with mass of 0.1g in the thermal neutron flux 10^{12}cm^{-2} s^{-1} for 1 hour. The cross section of auric activation by thermal neutrons is 97 barns

Solution:

Activity is called the number of disintegrations of this preparation in second. Activity equal disintegration probability decays on number nucleus radioactive isotope in the following form:

$J(t)=\lambda N(t)=In\sigma (1-e^{-\lambda t})$

With the condition, that exposure time is

$t<<T_{1/2}$
$\lambda t=t\cdot ln2/T_{1/2}<<1; (1-e^{-\lambda t})\approx 1-1+\lambda t$

Taking into account that $n=m\cdot N_a/A$, where m –mass of the activated model N_A - Avogadro's number, we obtain, what induced activity of the isotope of gold-198 comprises

$J=Im (N_A t\sigma / AT_{1/2})ln2 \approx 3,2.10^8Bq \approx 10mCi$

89. They are given the difference of the masses of atoms-
$\Delta(^{114}Cd_{48})$= -90.021 Mev, $\Delta(^{114}In_{49})$= -88.379 Mev and $\Delta(^{114}Sn_{50})$= -90.558 Mev

Determine the possible forms of β- nuclear decomposition of ($^{114}In_{49}$)

Solution:

For the nucleus of ^{114}In beta - disintegrations appear as follows

β^-- disintegration - ^{114}In$_{49}$ $\rightarrow$ ^{114}Sn$_{50}$ + e^-+ $\bar{v}_e$
β^+-disintegration - ^{114}In$_{49}$ $\rightarrow$ ^{114}Cd$_{48}$ + e^+ + v_e
e- capture - ^{114}In$_{49}$+ e^- $\rightarrow$ ^{114}Cd$_{48}$ + v_e

If the value of the decay energy is positive, the nucleus is unstable to the disintegration of this type. Energy of the disintegrations is:

β⁻-disitegration $\qquad$ $Q_\beta^- = \Delta(A,Z) - \Delta(A,Z+1)$
β⁺-disintegration $\qquad$ $Q_{\beta+} = \Delta(A,Z) - \Delta(A,Z-1) - 2m_ec^2$
e- capture $\qquad$ $Q_e = \Delta(A,Z) - \Delta(A,Z-1)$

Where $\Delta(A, Z)$ - the surplus of the masses of initial nucleus, $\Delta(A, Z+1)$ and $\Delta(A, Z-1)$ - the surpluses of the masses of residual nucleus, m_e - mass of the electro.:

Let us substitute the values:

β⁻decay $\qquad$ Q_β-= 90.558 - 88.379
$\qquad$ 2.179 Mev>0

β⁺-decay $\qquad$ $Q_{\beta+}$ = 90.021 - 88.379 - 1.022
$\qquad$ =0.62Mev>0

e- capture $\qquad$ Q_e= 90.021 - 88.379
$\qquad$ = 1.642Mev>0

Thus, nucleus $^{114}In_{49}$ experiences all three forms β- disintegration

90. For $^{17}Ne_{10}$ nucleus, determine the maximum energy of the being late protons, which depart from the $^{17}F_9$ nucleus, which is formed as a result of e- Capture on nucleus $^{17}Ne_{10}$

Solution:

The binding energies $E_{BE}(^{17}Ne)= 112.91$ Mev, $E_{BE}(^{17}F)= 128.23$ Mev and $E_{BE}(^{16}O)=126.63$ Mev

Consideration process:

$$^{17}Ne+ e^- \quad\rightarrow\quad {}^{17}F^*+ v_e \quad\rightarrow\quad {}^{16}O+ p$$

The maximum excitation energy of nucleus $^{17}F^*$ is equal to energy e-capture

$E_{max}(^{17}F^*)= Q_e$
$\quad =E_{BE}(^{17}F) - E_{BE.}(^{17}Ne) - 0.78$ Mev
$\quad =128.23\text{Mev}-112.91\text{Mev}$
$\quad =14.54\text{Mev}$

The separation energy of the proton for the ^{17}F nucleus is:

$\varepsilon_p= E_{BE}(A, Z) - E_{BE}(A\text{-}1, Z\text{-}1)$
$\quad =E_{BE}(^{17}F) - E_{BE}(^{16}O)$
$\quad =128.23\text{Mev}-126.63\text{Mev}$
$\varepsilon_p= 1.6$ Mev

The maximum energy of the being late protons T_p^{max} is

$T_p^{max}= \{E_{max}(^{17}F) - \varepsilon_p\}. M(^{16}O) /\{ M(^{16}O)+m_p\}$
$\quad \approx \{E_{max}(^{17}F) - \varepsilon_p\} 16/17$
$\quad \approx (14,54\text{Mev} - 1,6\text{Mev}) 16/17$
$\quad \approx 12,2\text{Mev}$

91. Find the separation energy of a neutron and of a proton from the nucleus of $^{12}C_6$?

Solution:

The separation energy of the neutron is:

$B_n = M(A - 1,Z) + m_n - M(A,Z)$
$\quad = \Delta (A - 1,Z) + \Delta_n - \Delta (A,Z)$
$B_n(^{12}C) = 10.650$ Mev $+ 8.071$ Mev $= 18.72$ Mev

The separation energy of the proton is:

$B_p = M(A-1,Z-1) + M(^1H) - M(A,Z) = \Delta (A-1,Z-1) + \Delta (^1H) - \Delta (A,Z)$
$B_p = (^{12}C) = 8.668$ Mev $+ 7.289$ Mev
$B_p = 15.96$ Mev

92. How many times is the number of nuclear decompositions of radioactive iodine $^{131}I_{53}$ during the first day more than the number of disintegrations during the second day? The period of the half-life of $^{131}I_{53}$ isotope equal to 193 hours.

Solution:

From the radioactive decay law $N(t) = N_0 e^{-\lambda t}$ follows that during the first day (first 24 hour) it was decomposed:

$N_1 = N_0(1 - e^{-24\lambda})$ nuclei

During the second day it was decomposed

$N_2 = N_0(1 - e^{-24\lambda})e^{-24\lambda}$ nuclei. Ratio of the number of disintegrations in the first day

$N_1/N_2 = e^{24\lambda} = e^{24 \ln 2 / T1/2}$ to the number of disintegrations in the second day

Where $T_{1/2}$- the period of the half-life of ^{131}I in the hours, connected with λ relationship $T_{1/2} = \ln2/\lambda = 0.693/\lambda$

It is final.

$N_1/N_2 = e^{24.0,693/T1/2} = e^{24.\ 0,693/193}$
$N_1/N_2 = 1,09$

93. Determine the isolated energy W, by 1 mg of preparation of $^{210}Po_{84}$ in the time which equal to the average of the life, if the energy E is 5.4Mev separated with one report of the disintegration.

Solution:

The number of the nuclei of radioactive preparation in the average of the life decreases in e= 2.718 times.

Then a quantity of decomposed nuclei in this time will be:

D= 1 - 1/2.718= 0.632 from their prime number

The initial number of nuclei N in the model with a mass of mg is determined from the relationship of:

$N = mN_A/A$

Where N_A-Avogadro's number, A - mass number

Quantity of energy, which was isolated in the time, equal to the average of the life of ^{210}Po isotope:

$W = DmN_A E / A$
$= \{0,632.10^{-3}g.\ 6,02.10^{23}mol^{-1}.\ 5,4Mev.\ 1,6.10^{-6}erg/Mev\}/210g/mol$
$W = 1,6.10^{13}erg$

94. Determine upper boundary of the age of the Earth, considering that entire existing on the earth, $^{40}Ar_{18}$ was formed from $^{40}K_{19}$ as a result of e-capture. At present to every 300 atoms of $^{40}Ar_{18}$ is fallen one atom $^{40}K_{19}$

Solution:

The number of ^{40}K nuclei undecomposed at the present time

$N_K = N_{Ar} / 300 = N_0 e^{-t \ln2/T1/2}$

Where N_0 - initial number of ^{40}K nuclei at the moment of the earth formation

t - The age of the earth
$T_{1/2}$ - period of the half-life of ^{40}K which is $1.277 \cdot 10^9$ years

With radioactive decay of ^{40}K by e seizure are decomposed only 10.67% of nuclei; therefore the number of argon nuclei will be at the present time:

$N_{Ar} = 0,1067 N_0 (1 - e^{-t\ln2/T1/2})$

We obtain the equation:

$300 N_0 e^{-t\ln2/T1/2} = 0,1067 N_0 (1 - e^{-t\ln2/T1/2})$

From where:

$t = -\ln[0,1067/300 + 0,1067]\{1,277.10^9 y/\ln2\}$
$t \approx 1,5.10^{10}$ year

95. As a result of α- disintegration radium $^{226}Ra_{88}$ is converted into radon $^{222}Rn_{86}$. What is the volume of radon with the standard conditions which located in the equilibrium with 1 g of radium? The period of half-life $^{226}Ra_{88}$; $T_{1/2}$ is 1600 years, and for $^{222}Rn_{86}$ is $T_{1/2} = 3.82$ days

Solution:

In the establishment of the secular equilibrium a quantity of radioactive nuclei of both isotopes and their disintegration constants are connected with equation of: $\lambda_1 N_1 = \lambda_2 N_2$

From where:

$$N_{Rn} = N_{Ra}\lambda_{Ra}/\lambda_{Rn} = N_{Ra}T_{1/2}(Rn)/T_{1/2}(Ra)$$

The quantity of ^{226}Ra nuclei is:

$$N_{Ra} = m\, N_A/A$$

Where m and A the mass and the mass number of ^{226}Ra

N_A - Avogadro's number

The desired volume is:

$$V = V_M N_{Rn}/N_A$$

Where V_M - the molar volume of gas (22.4 l/mole). We obtain

$$V = V_M m T_{1/2}(Rn)\, /\, A T_{1/2}(Ra)$$
$$= [\ 22,4\ell/mol\ .\ 1g\ .\ 3{,}82 days]\ /\ [226 g/mol.\ 1600 year.365 day/year]$$
$$V = 6{,}5.10^{-7}\ell eter$$

96. A proton collide a target of the natural Lithium. When the activity was 140Bq, which the radioactive source of this activity, when the time was 106 days to reduce the activity to 35Bq?

Solution:

The relation of the activity is:

$$A = A_0\, e^{-\lambda t}$$

Then:

$A / A_0 = e^{-\lambda t}$

And $\ln(A / A_0) = -\lambda t$

Therefore:

$\lambda = -\{\ln(A/A_0) /t\}$

When:

$T_{1/2} = \ln 2 / \lambda$
$T_{1/2} = \ln 2 / -\{\ln(A/A_0)/ t\}$
$\qquad$ t. $\ln 2 / \ln(A/A_0)$

And:

$T_{1/2} = 106. \, 0,693 / \ln(140/35)$
$T_{1/2} = 73,458 / 1,38629$
$T_{1/2} = 52,988$ days

Therefore:

$T_{1/2} \approx 53$ days

Therefore the radioactive source is Beryllium.

97. A plate from ^{55}Mn with a thickness d= 0.2cm, which radiated the beam of neutrons J in the time of t_{act}= 12 min, if t_{oxl}= 100 min and the activity was 2800 Bq, Calculate the intensity of neutron beam J, when the cross section was σ= 0.48 barns, and the density of the plate is ρ= 7.43 g/cm³.

Solution:

We calculate the decay constant:

$\lambda = 0{,}693 / T_{1/2}$
   $= 0{,}693 / 2{,}5785.60$
$\lambda = 4{,}479.10^{-3} \ min^{-1}$

The activity of the plate is:

$I = Jn\ \sigma(1-e^{-\lambda tact})e^{-\lambda tox}$
And: $J = I / n\ \sigma(1-e^{-\lambda tact})e^{-\lambda tox}$
Where n - number of nuclei per unit of the area of target.
$n = \rho dN_A / A$

Hence:

$J = IAe^{\lambda tox} / \rho dN_A\ \sigma(1-e^{-\lambda tac})$
   $= [2100.\ 55\ e^{-(4,479.10-3).100}] / [7{,}43.0{,}2.6{,}025.10^{23}.\ 0{,}48.10^{-24}(1-e^{-(4,479.10-3).15}]$
   $= 98401{,}178 / \{(0{,}4297512).\ (0{,}0649775)\}\ )$
   $= 98401{,}178 / 0{,}0279241$
J= 3523880 neutron / min

98. How much energy will be released when two deuterons. Combine to form an alpha-particle?

Solution:

The energy which released= $(2.2{,}014102) - (4{,}002602)$

When the mass of $^2D_1$ is: 2,014102amuand of alpha is 4,002602.

And $Q = 4{,}028204 - 4{,}002602$
      $= 0{,}025602amu$
Or $Q = 0{,}025602.\ 931{,}5$
     $Q = 23{,}85Mev.$

99. The isotope of $^{218}Po_{84}$ can decay by either α or β emission. What is the energy release in each case? The mass of $^{218}Po_{84}$ is 218,008965amu

Solution:

If this decay is by alpha particle:

Hence:

$$^{218}Po_{84} \rightarrow {}^{214}Y_{82} + {}^{4}He_2 + Q$$

The result nuclie is $^{214}Pb_{82}$,

Hence:

$\Delta m = 218,008965 - (213,999798 + 4,002602)$
$\Delta m = 218,008965 - 218,00239$
$\Delta m = 6,57. 10^{-3}amu$

Hence the energy is $6,57. 10^{-3} .931,5 = 6,119955$Mev.

And if this decay is by β⁻,

$$^{218}Po_{84} \rightarrow {}^{218}Y_{85} + \beta + Q$$

Hence the release energy is:

$\Delta m = 218,008965 - (218,00868: {}^{218}At_{85} + 5,4859317. 10^{-4})$

Hence:

$\Delta m = 218,008965 - 218,00922 = -2,6 . 10^{-4}$ amu

Hence Q = - 0,24219Mev.

And Q = - 242,19 Kev.

That means: this decay need 242,19 Kev
In order to make, because the sign is negative.

100. The nuclide of $^{32}P_{15}$ decay by emitting an electron whose maximum kinetic energy can be 1,71Mev.

    (a) What is the daughter nucleus?
    (b) What is its atomic mass in (amu)?

Solution:

$$^{32}P_{15} \longrightarrow {}^{A}Y_{Z} + \beta^{-}$$
$$1,71Mev$$

When the daughter nucleus has Z= 16 and A=32

Hence $^{A}Y_{Z}$ is $^{32}S_{16}$

And $^{32}P_{15} \longrightarrow {}^{32}S_{16} + \beta^{-}$

And atomic mass unit is:

$$(15m_{p}+ 17m_{n}) \longrightarrow 16(m_{p}+ m_{n})+ 5,4859317. \, 10^{-4}$$
31,973908amu   30,966083amu

When amu for $^{32}P_{15}$ is 31,973908amu

And amu for β^{-} is 5,4859317. 10^{-4} amu

Hence the daughter amu is 31,973908 - 5,4859317. 10^{-4} = 31,97336amu

101. Show that the decay $^{11}C_{6} \longrightarrow {}^{10}B_{5}+ {}^{1}P_{1}$ is not possible because the energy would not be conserved.

Solution:

The first: we find the B.E. of $^{11}C_{6}$:

The protons number of this nucleus is 6 and the neutrons number of this nucleus is 5,

Hence the number of the particles in this nucleus becomes:

$6m_p + 5m_n = 6(1,007825) + 5(1,008665)$
$$= 6,04695 + 5,043325$$
$$= 11,090275 \text{amu}$$

And M_{11C6} (the mass of nuclied) from the appendix is: $11,011433$amu

Hence binding energy= $(11,090275 - 11,011433). 931,5$
$$= 0,078842 \text{amu}. 931,5$$
$$E_B = 73,441323 \text{Mev}.$$

The second step is:

We find the binding energy of the production:

The protons number of this nucleus is 5 and the neutrons number of this nucleus is 5 too.

Hence:

$(5m_p + 5m_n) = 5(1,007825 + 1,008665)$
$$= 10,08245 \text{amu}$$
And $M_{10B5} = 10,012936$amu

$\Delta m = 10,08245 - 10,012936 = 0,069514$amu

Hence the binding energy of $^{10}B_5$ is:

$E_B = 0,069514 . 931,5 = 64,752291$Mev.

And the energy of the protons is: $1,007825 . 931,5 = 938,78898$Mev.

The energy of the production is:

64,752291+ 938,78898 = 1003,5412Mev.

Hence:

$E_{11C6} < E_{prod}$.(64,75 < 1003,54).

102. a) Show that the nucleus of $^8Be_4$ (m= 8,005308amu) is unstable to decay into two α particles.

(b) Is $^{12}C_6$ stable against decay into three α particles? Show why or why not.

Solution:

(a) $4(m_p+ m_n) =$
$$= 4(1,007825+ 1,008665)$$
$$= 8,06596amu$$
$M_{Be} = 8,005308amu$

Hence $\Delta m= 0,060652amu$

And the binding energy = 0,060652 . 931,5
$$= 56,497338Mev.$$
$2 (^4He_2) = 2 (4,002602amu)$
$$= 8,005204amu$$

And $2(2m_p+ 2m_n)=$
$$= 4,03298 .2$$
$$= 8,06596amu$$

And:

$E_B= 8,06596 – 8,005204$
$= 0,060756amu$
$E_B = 56,594214Mev.$

Hence:

Because of this energy (of two α) > energy of $^8Be_4$.

(b) $6(m_p + m_n) =$
$$= 6(1,007825 + 1,008665)$$
$$= 12,09894 amu$$
$M_{12C6} = 12$

Hence $\Delta m = 12,09894 - 12$
$$= 0,09894 amu$$

And:

$E_B = 92,16261 Mev.$
$3\,(^4He_2) = 3(4,002602 amu)$
$$= 12,007806 amu$$
And $2(m_p + m_n) = 4,03298 amu$

Hence:

E_B of $3\alpha = 3(4,03298)$
$$= 12,09894 amu$$
$$= 12,09894 - 12,007806$$
$$= 0,091134 amu$$
$$= 84,891321 Mev.$$

We can say:

Yes that is right because the energy of $^{12}C_6$ > energy of 3α.

103. *The maximum permissible concentration of radium emanation in the air for continuous exposure is 10^{-8} micro curies per milliliter of the air. For this concentration what is the radon content of 1 ml of standard air in percent by the mass and in percent by the* *volume? The density of the air at standard conditions is 1,2929 . 10^{-3} gm /cm^3*

Solution:

We must find the mass and the volume for 10^{-8} micro curies to the radon in the standard conditions:

$\lambda_{Rn} = 2,1.\ 10^{-6}\ sec^{-1}$

Now:

$10^{-8}\ \mu Ci = 10^{-8}.\ 10^{-6} = 10^{-14}$ Curie.

For one curie= $3,7.\ 10^{10}$ dis/sec

Hence:

$10^{-8}\ \mu Ci = 10^{-14}.\ 3,7.\ 10^{10}$
$= 3,7.\ 10^{-4}$ dis / sec.

Hence:

$3,7.10^{-4} = \lambda N$
$= 2,1.\ 10^{-6}.\ 6,025.\ (10^{23} / 222).\ M\ (gm).$

Hence:

$M = 3,7.\ 10^{-4}.222 / 2,1.10^{-6}.6,025.10^{23}$
$= (3,7.222) / (2,1.6,025) .10^{-21}gm$
$M = 65.\ 10^{-21}$ gm of the radon...................(1)

The volume of this mass in (1) in standard conditions is:

$65.10^{-21} / 222 . 22400$
$\approx 65.10^{-19}cm^3$......................(2)

In 1,2929 . 10^{-3} gm /cm^3 of the air be found 65. 10^{-21} gm of the radon

Hence the percentage is 5. 10^{-15} %

If we translate one liter to cubic centimeter:

1liter= $1000cm^3$
ml= $1cm^3$.

In each one milliliter of the air be found $65.10^{-19}cm^3$ from the radon or

6,5. 10^{-16}%.

104. What is the weight in grams of 1 curie of 5,3 year of $^{137}Cs_{55}$?

Solution:

Let N is the number of atoms of ^3T equivalent to one curie.

$T_{1/2}(^3T)$ = 12,33. 365,25. 24. 3600
$\qquad$ = 3,89 . 10^8 sec
-dN /dt = λN
$\qquad$ = 3,7.10^{10} dis/sec
$\qquad \lambda$ = 0,693 / 3,89. 10^8 = 1,78.10^{-9} sec^{-1}
N $\qquad$ = 3,7. 10^{10} / 1,78. 10^{-9} = 2,077. 10^{19}
Weight of ^3T= (2,077.10^{19} . 3) / (6,025. 10^{23})
Weight of ^3T= 1,034. 10^{-4} gm.

105. What is the activity of a gold foil that has been irradiated for a period of 10 hours, assuming a constant rate of the formation of the radioisotope in the reactor of 10^9 atoms per sec, if its half-half is 53,3 days.

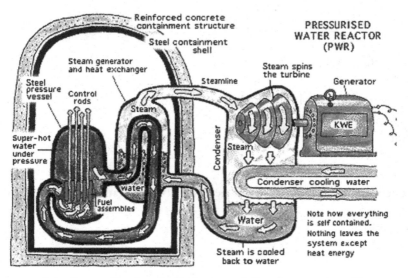

http://www.freedomforfission.org.uk/img/pwrreactor.gif

Solution:

$\lambda = \ln2 \, /T$

$\quad = 0{,}693 \, / \, 53{,}3.24h$

$A = \lambda N = R \, (\, 1\text{-}e^{-\lambda t})$

$\quad = 10^9 \, (\, 1\text{-} \, e^{-0{,}693.10 \, / \, 53{,}3. \, 24})$

$\quad = 10^9 \, (\, 1\text{-} \, e^{-5{,}417..0{,}001})$

$\quad = 10^9 \, (1\text{-} \, 0{,}9946)$

$\quad = 5{,}4.10^6 \; dis/sec$

$\quad = 5{,}4.10^6 \, / \, 3{,}7.10^{10} \; curies$

$A = 0{,}146 \; mCi$

106. How much energy is required to remove a neutron from $^{236}U_{92}$ nucleus?

Solution:

The mass of $^{236}U_{92}$ is 236,045562amu and the remaining nucleus is $^{235}U_{92}$ whose have mass is 235,043924amu the mass of $^1n_0$ is 1,008665amu Therefor the mass of the final system is:

235,043924+ 1,008665= 236,05258amu

Hence the mass increase of the system is:

236,05258 – 236,045562= 7,02.10^{-3}amu

Thus the energy required to remove a neutron from the nucleas of $^{236}U_{92}$ is:

7,02.10^{-3} . 931,5 = 6,54Mev.

107. *A sample has a $^{14}C_6$. The activity of 0,0061 Bq per gram of carbon. Find the age of the sample, assuming that the activity pergram of carbon in a living organism has been constant at a value of 0,23 Bq.*

Solution:

$\lambda = 0,693 / T_{1/2}$
   $= 0,693 / 5730$
   $= 1,209. 10^{-4}$ yr^{-1}

$A = A_0 e^{-\lambda t}$
0,0061 $= 0,23$ e$^{-(1,209.10-4) t}$
0,0061 / 0,23 $=$ e$^{-(1,209.10-4) t}$
0,0265217$=$ e$^{-(1,209.10-4) t}$
ln 0,0265217$= -1,209.10^{-4t}$
-3,62979$= -1,209.10^{-4}$t

Hence:

t = 30023,076 years.

108. *The practical limit to ages that can be determined by radiocarbon dating is about 41000yr. In a 41000 yr-old sample, what percentage of the original $^{14}C_6$ atoms remains?*

Solution:

$\lambda = 0{,}693 \, / \, T_{1/2}$
  $= 0{,}693 \, / \, 5730 \text{ yr}$
  $= 1{,}209. \, 10^{-4} \text{ yr}^{-1}$

$A \, / \, A_0 = e^{-\lambda t}$
  $= e^{-(1{,}20942.10{-}4)(41000\text{yr})}$
  $= e^{-4{,}9586384}$

$A \, / \, A_0 = 7{,}02248.10^{-3}$
$A \, / \, A_0 = 0{,}007 \, \%.$

109. Bones of the woolly mammoth have been found in North America. The youngest of these bones has a $^{14}C_6$ activity per gram of carbon that is about 21% of what was present in the live animal. How long ago(in years) did this animal disappear from North America?

http://www.science.psu.edu/alert/images/SchusterMiller_
MammothBones.jpg

Solution:

$\lambda = 0,693 / T_{1/2}$
  $= 0,693 / 5730$ yr
$\lambda = 1,209. 10^{-4}$ yr^{-1}

$A / A_0 = e^{-\lambda t}$
$0,21 = e^{-(1,20942.10-4)(t)}$
$\ln 0,21 = -(1,20942.10^{-4})(t)$
$t = \ln 0,21 / -1,209.10^{-4}$
$t = 12908,577$ yr.

110. What is the mass (in grams) of krypton $^{92}Kr_{36}$ ($T_{1/2}= 1,84$ s) which has the same activity as 2gram of Xenon $^{140}Xe_{54}$ ($T_{1/2}=13,6$ s)?

Solution:

$\lambda_{xe} = 0,693 / 13,6$
$\lambda_{xe} = 0,051$ sec^{-1}
$N_{xe} / m_{xe} = N_A / A_{xe}$
$N_{xe} / 2g = 6,025.10^{23} / 140$

Hence:

$N_{xe} = (2)(6,025.10^{23}) / 140$
$N_{xe} = 8,607. 10^{21}$ nuclei.
$A_{xe} = |\Delta N / \Delta t| = \lambda_{xe} N_{xe}$
  $= (0,051). (8,607. 10^{21})$
$A_{xe} = 4,3858.10^{20}$ Bq (dis/sec)

And:

$\lambda_{kr} = 0,693 / 1,84$
$\lambda_{kr} = 0,3766$ sec^{-1}

When: $A_{xe} = A_{kr}$

Hence:

$A_{kr} = 4,3858.10^{20}$,

And:

$4,3858.10^{20} = (0,3766) N_{kr}$.

Hence:

$N_{kr} = 1,1645.10^{21}$ nuclei.

When $N_{kr} / m_{kr} = N_A / A_{kr}$

Hence:

$1,1645.10^{21} / m_{kr} = 6,025.10^{23} / 92$

Hence:

$m = (1,1645.10^{21})(92) / 6,025.10^{23}$

Hence:

$m = 0,1778$ gram.

111. *To see why one curie of activity was chosen to be 3,7.10^{10} Bq.*
Determine the activity of one gram of radium $^{226}Ra_{88}$ ($T_{1/2}=$
1,6.10^{3} yr).

Solution:

$A = A_0 e^{-\lambda t}$
$\lambda = 0,693 / T_{1/2}$
$\quad = 0,693 / 1,6.10^{3} \cdot 365,25. \, 24.3600$
$\quad = 1,3725.10^{-11} \text{ sec}^{-1}$

N_0 / mass $= N_A$ / A
N_0 / 1gram $= 6,025.10^{23}$ / 226
$N_0 = 6,025.10^{23}$ / 226
$N_0 = 2,666.10^{21}$ nuclei.

When: $A_0 = \lambda N_0$

Hence:

$A_0 = (1,3725.10^{-11})(2,666.10^{21})$
$\quad = 3,66.10^{10}$ dis / sec
$A_0 \approx 3,7.10^{10}$ Bq.

112. What is the rest energy with Mev /c^2 of an α-particle.

Solution:

$E_0 = 1/2\ m_0\ c^2 = \frac{1}{2}\ (2m_p + 2m_n)c^2$
$\qquad = \frac{1}{2}\ (2.\ 1,6726231.\ 10^{-27}kg + 2.\ 1,6749286.\ 10^{-27}kg)$
$\quad m_p c^2 = 938,272004 Mev$
$\quad m_n c^2 = 1,6749271.\ 938,272004\ /\ 1,67262161$
$\quad m_n c^2 = 939,56523 Mev$
$\quad E_0\ /\ c^2 = \frac{1}{2}.\ 2\ .\ 938,272004 + \frac{1}{2}.\ 2$
$\quad E_0\ /\ c^2 = 939,27004 Mev/c^2$

113. Calculate the decay energy of uranium nucleus($^{232}U_{92}$) which decay to thorium $^{228}Th_{90}$ and alpha particle.

Solution:

The atomic mass of an alpha particle ($^4He_2$)= 4,002602amu

The total mass in the final case is:

228,028716amu+ 4,002602amu= 232.031318amu

The lost mass when the nucleus $^{232}U_{92}$ decay is:

232,037131amu – 232,031318amu= 0,005813amu

When 1amu= 931,5Mev. So the released energy Q in this reaction is:

Q= (0,005813)(931,5Mev/amu)
Q ≈ 5,4Mev.

This energy will appear as a kinetic energy to an alpha particle and the daughter nucleus. We can say that we can apply the conversation of momentum to appear a kinetic energy to an alpha particle in this decay which equal to 5,3Mev and to the daughter nucleus which equal to 0,1Mev, and which take the inverse direction of alpha particle.

114. The 2 microgram of Radium nucleus ($^{A}Ra_{88}$) which has the atomic number of 88. Its nucleus irradiates an alpha particle and follows gamma rays. The half-life of radium is 1600 years. Calculate the number of alpha particles in the decay process

Solution:

λ = 0,693 / $T_{1/2}$
    = 0,693 / 1600. 365,25. 24.60.60
    = 1,37. 10^{-11} sec^{-1}

The number of atoms in 2 □gram to radium is:

N= (6,025. 10^{23} / 226). 2. 10^{-6} gm
N= 5,332 . 10^{15} atoms.

The number of decays in one second= The number of alpha particles which `emitted in one second.

$A = N \lambda$
$= 5,332 . 10^{15} . 1,37. 10^{-11}$
$A \approx 73046$ alpha particles / sec.

115. What is the mass and the volume under standard conditions of two curie of radium emanation (radon)?

Solution:

$\lambda = 0,693 / (3,8235. 86400)$ sec^{-1}
$= 2,0977.10^{-6} \approx 2,1.10^{-6}$ sec^{-1}
$N= (6,025.10^{23} / 222)$ M

Hence the activity $(A)= N\lambda = (6,025. 2,1 / 222). 10^{17}$ M dis/sec
$M = (3,7.10^{10}. 222) / (6,025. 2,1.10^{17})$
$M = (3,7.222 / 6,025. 2,1). 10^{-7}$ gm.

The number of molars in radon which has the mass of M/222.

Hence this number has a volume in the standards conditions of:

$M/222. 22400$ m^3(standard volume)$= (3,7. 224/ 6,024. 2,1). 10^{-5}$cm^3
$= 0,655$ mm^3.

116. What the ratio at the time of the earth formation when the ratio of the number of $^{238}U_{92}$(the half-life is $4,5.10^9$ years) and the number of $^{235}U_{92}$(the half-life is $7,1.10^8$ years) when this ratio between the two isotopes is $140(^{238}U_{92} / ^{235}U_{92} = 140)$ and the earth is 4.10^9 years old.

Solution:

$\lambda_{238} = 0,693 / 4,468. 10^9$
$\lambda_{235} = 0,693 / 7,038. 10^8$
$N_{238} / N_{235} = (N_0)_{238}. e^{-\lambda_1 t} / (N_0)_{235}. e^{-\lambda_2 t}$

Hence $140 /1 = (N_0)_1 / (N_0)_2 . e^{(\lambda_2 - \lambda_1) t}$

When: $t = 4. 10^9$ year

Hence:

$(\lambda_2 - \lambda_1) t = (0,693 / 7,038. 10^8 - 0,693 / 4,468.10^9). 4. 10^9$
$\qquad = (9,846.10^{-10} - 1,551. 10^{-10}). 4.10^9$
$(\lambda_2 - \lambda_1) t = 3,318$
$N_1 / N_2 = 140 \, e^{-3,318}$
$N_1 / N_2 \approx 5,07.$

117. A sample of phosphorus contains 8% atoms of phosphorus ($^{32}P_{15}$) with a the radioactivity of 0,023 micro curie.

   a) Find the number of atoms of this isotope in this sample.
   b) Find the total mass of this sample($T_{1/2}$= 14,262days).

Solution:

a) $\lambda = 0,693 / T_{1/2}$
$\qquad = 0,693 / 14,262.24.3600$
$\quad \lambda = 5,624.10^{-7} \, sec^{-1}$
$A = \lambda N$

Hence:

$0,023.10^{-6}.3,7.10^{10} = 5,624.10^{-7} . N$
$N = 0,0851.10^4 / 5,624.10^{-7}$
$N = 1,513.10^9$ nuclei.

(8/ 100). $(1,513.10^9)= 1,21.10^8$ atoms of phosphorus in the sample.

b) The total mass of this sample is:

$N / m = N_A / A$
$1,513.10^9 / m = 6,025.10^{23}/32$

Therefore:

$m = 1,513.10^9.32/6,025.10^{23}$
$m = 8,036.10^{-14}$gram.

118. How many alpha particles are emitted per second by 3 microgram of radium?

Solution:

The half-life of $^{226}Ra_{88}$ is:

$T_{1/2}= 1600$ years.
$\lambda = 0,693 / 1600. 365.25. 24. 3600$
$\lambda = 1,372.10^{-11}$ sec^{-1}

The number of atoms in 3 microgram of $^{226}Ra_{88}$ is:

$N = (6,025.10^{23} .3. 10^{-6}) / 226$ gm$= (6,025. 10^7 / 226)$ atoms

Therefore:

The number of disintegrations per sec is:

$N\lambda= 1,372.10^{-11}$ sec$^{-1.}$ 6,025 /226. 10^7
$= 1,372. 6,025.10^6 / 226$
$N\lambda \approx 36577$alpha particles / sec.

119. How many disintegrations occur each second in 0,5gram of $^{243}Am_{95}$, and what is the number of curies in this isotope ?

Solution:

The half life of this isotope is: 7380years,

λ = 0,693 / 7380. 365,25. 24. 3600
  = 0,693 / 2,329.10^{11}
λ = 2,9756. 10^{-12}sec^{-1}

The number of atoms in 0,5 gm of this isotope is:

N = 6,025.10^{23}.0,5 / 243
N = 1,2397.10^{21} atom
Activity(A)= λ N
  = 2,9756.10^{-12}. 1,2397.10^{21}
A = 3,6888.10^9 dis/sec.

And we know that each one curie= 3,7.10^{10} dis/sec

Therefore:

A = 3,6888.10^9 / 3,7.10^{10}
A= 0,0997 curies.

120.   What is the energy of an alpha particle which released in the following disintegration?

$$^{210}Po_{84} \rightarrow\ ^{206}Pb_{82} + \alpha + Q(?)$$

Solution:

$Q_\alpha = (M_{po} - M_{pb})c^2$
  = (209,982848amu - 205,97444amu)c^2
  = (4,0084amu) c^2 . 931,5
Q_α = 3733,8246Mev.

121. In 9 days the number of radioactive nuclei decreases to one-eight the number present initially. What is the half-life (in days) of the material?

Solution:

$T_{1/2} = \ln 2 / \lambda$

Since $N = N_0\, e^{-\lambda t}$

When: $N = (1/8)\, N_0$

Hence:

$(1/8)\, N_0 = N_0\, e^{-\lambda(9)}$
$0,125 = e^{-\lambda(9)}$

Hence:

$-2,079441 = -9\,\lambda$

And:

$\lambda = -2,079441\,/\,-9$
$\lambda = 0,23105\ \text{days}^{-1}$
$T_{1/2} = 0,693/0,23105$
$T_{1/2} = 3\ \text{days.}$

122. The number of the radioactive nuclei present at the start of an experiment is $4,6.10^{15}$. The number present twenty days later is $8,14.10^{14}$. What is the half-life (in days) of the nuclei?

Solution:

$N = N_0\, e^{-\lambda t}$
$8,14.10^{14} = 4,6.10^{15}\ e^{-\lambda(20\text{days})}$

$0,1769565 = e^{-\lambda(20)}$
$-1,731851 = -20\,\lambda$

Hence:

$\lambda = 0,0865925\ \text{day}^{-1}$
$T_{1/2} = 0,693 / 0,0865925$
$T_{1/2} = 8\ \text{days}.$

123. A radioactive material decay of 1280 dis/minute. After 6 hours its decay becomes 320 dis/minute. What is the half-life for this material?

Solution:

$A = A_0 \cdot e^{-\lambda t}$
$320 / 3600 = \{ 1280 / 3600 \} \cdot e^{-\lambda(21600)}$
$0,249999 = -\lambda.\ 21600$

Therefor:

$\lambda = 6,4180277.10^{-5}\ \text{sec}^{-1}$

When:

$T_{1/2} = \ln2 / \lambda = \ln2 / 6,4180277.10^{-5}$
$T_{1/2} = 180\ \text{min}.$

124. What is the decay constant for Uranium($^{238}U_{92}$) which has a half-life of $4,468.10^{9}$ year?

Solution:

$T_{1/2} = \ln / \lambda$

4,468.10⁹ = 0,6931471 / λ

$\lambda = 1,5513587.10^{-10}$ yr⁻¹

125. What is the activity for $^{14}C_6$ nucleus which has $5,6.10^{20}$ nuclei?

Solution:

When : $N = N_0 e^{-\lambda t}$; and: t= 0 ; N= $N_0 . e^0 = N_0$

At t= $T_{1/2}$;
N= $N_0 / 2$ by the definition of $T_{1/2}$
$N_0 / 2 = N_0 e^{-\lambda\ T1/2}$,
$1/2 = e^{-\lambda\ T1/2}$,

Or: $e^{\lambda\ T1/2} = 2$

$\ln(e^{\lambda\ T1/2}) = \lambda T_{1/2} = \ln2$,
$T_{1/2} = \ln2/\lambda = 0,693 / \lambda$
$\lambda = 0,693 / T_{1/2}$
= 0,693 / {(5730yr)(3,156.10⁷s/yr)}
$\lambda = 3,83.10^{-12}$ s⁻¹,

Hence:

A = |ΔN / Δt |= λN
A = (3,83. 10⁻¹² s⁻¹)(5,6.10²⁰)

Therefor:

A = 2,1448.10⁹ decay /sec.

126. The isotope $^{243}Am_{95}$ has a half-life of 7380 y.
   a) How many nucleus in this isotope, when its activity is $5,5.10^{15}$ decay/sec.

135

b) **What is the activity of this isotope after 500y and**
c) **What is the activity after 7000y?**

Solution:

$\lambda = 0,693 / (7380. 365,25. 24. 3600sec)$
   $= 2,9755.10^{-12} sec^{-1}$

The activity $A = \Delta N/\Delta t = \lambda N$

$2,9755.10^{-12} sec^{-1} . N = 5,5.10^{15}$ decay / sec

Hence:

$N = 1,848.10^{27}$ nuclei.

b) after 500y its activity will be:

$A = A_0 . e^{-\lambda t}$
$A_{500y} = 5,5.10^{15}. e^{-2,9755.10-12(500. 365,25.24.3600sec)}$
$A_{500y} = 5,5.10^{15} . e^{-0,0469512}$
$A_{500y} = 5,5.10^{15} . 0,9541341$
$A_{500y} = 5,247.10^{15}$ decay/sec

c) And in the same method we can find the activity after 7000 y which
   is of

$A_{7000y} = 2,85.10^{15}$ decay/sec.

We can see that the activity arrive approximately to the half of the
initial activity

127. **$2\mu g$ of $^{235}U_{92}$, which has a half-life of $7,038.10^8 y$.**
   a) **How many nuclei in this nucleus?**
   b) **What is the initial activity for this isotope?**
   c) **What is the activity after 5 million years?**

d) Which the time must be the activity 1/10 of its initial activity?

Solution:

a) When the atomic mass for this isotope is: 235,043924amu

Then:

235grams will be: $6,02.10^{23}$ nuclei (Avogadro number).

Hence, it has 2.10^{-6}gram which include initial nuclei number as:

$N_0 / 2.10^{-6} = 6,02.10^{23} / 235g$

Hence:

$N_0 = 5,123.10^{15}$ nuclei

b) From the equation as: $T_{1/2} = 0,693 / \lambda$

Hence:

$\lambda = 0,693 / 7,038.10^8.365,25.24.3600$ sec
$\lambda = 3,1201826.10^{-17}$ sec^{-1}

When: t= 0

Hence: $(\Delta N / \Delta t)_0 = \lambda N_0$
$\qquad = 3,12018.10^{-17}$ sec^{-1}. $5,1234.10^{15}$ nuclei
$\qquad = 0,1598595$ decay /sec

c) When the half-life is: $7,038.10^8$ y

Hence the decay constant decrease to the half after the same time.

d) $(\Delta N / \Delta t) = (\Delta N / \Delta t)_0. e^{-\lambda t}$

$0,1 (5044784,1) = 5044784,1$ day/year . $e^{-9,8465473.10-10\,t}$

Then:

$0,1 = e^{-9,8465473.10-10\,t}$

Hence:

$\ln 0,1 = -9,84654.10^{-10}\,t$
$-2,302585 = -9,84654\,t$

Hence:

$t = 2,33846.10^9$ years.

128. The boney form has 9,2grams of $^{14}C_6$. Its radioactivity is 1,6Bq. How many years old this boney form?

Solution:

$N_0 / 9,2g = 6,024.10^{23} / 14$

Therefor:

$N_0 = 3,9586.10^{23}$ nucleus.
$(\Delta N/\Delta t) = \lambda N_0$
$1,6\,Bq = \lambda(3,9586.10^{23}\text{nucleus})$
$\lambda = 4,0418.10^{-24}\,sec^{-1}$
$T_{1/2} = 0,693 / \lambda$
$T_{1/2} = 1,7145.10^{23} / \{365,25.24.3600\}sec$

Hence:

$T_{1/2} = 1,72167.10^8$ years.

129. How many nucleus of $^{238}U_{92}$ in the stone , when the activity of this isotope is of $1,8.10^4$ decay / sec?

Solution:

The half-life of $^{238}U_{92}$ is: $4,468.10^9$ years

Then:

$\lambda = 0,693 / \{4,468.10^9. 365,25. 24. 3600 \ sec\}$
$\lambda = 4,9149162.10^{-18} \ sec^{-1}$
$A = \lambda N$

And:

$N = A / \lambda$
$\quad = \{1,8.104 \ decay \ /sec\} / 4,9149162.10^{-18} \ sec^{-1}$
$N = 3,6623.10^{21}$ decay per second.

130. The nuclide $^{124}Cs_{55}$ has half-life of 30,8 sec.
    a) If its 9,5 microgram in the initial state , how many nucleons in this state?
    b) How many nucleons will be after 2 minutes?
    c) What is the activity at this time?
    d) After which the time decreases the activity to the less than one in the second?

Solution:

$T_{1/2}$ is 30,8 second.

Therefor:

a)

$\lambda = 0,693 / T_{1/2}$

$\lambda = 0,693 / 30,8 = 0,0225 \text{ sec}^{-1}$

$N_0 / 9,5.10^{-6} \text{ g} = 6,024.10^{23} / 124$

Hence:

$N_0 = 4,612096.10^{16}$ nuclei

$(\Delta N/\Delta t) = \lambda N_0$

$\quad\quad = 0,0225. \ 4,612.10^{16}$

$(\Delta N/\Delta t) = 1,03772.10^{15}$ decay/sec

b) $N = N_0 \ e^{-\lambda t}$

$\quad N = 4,6121.10^{16}. \ e^{-0,0225. \ 2. \ 60}$

$\quad N = 3,0995.10^{15}$ nuclei

c) $A = \lambda N$

$\quad\quad = 0,0225. \ 3,0995.10^{15}$

$\quad A = 6,974. \ 10^{13}$ decay/sec

d) t= ?

$\quad N = N_0 \ e^{-\lambda t}$

$\quad 3,0995.10^{15} = 4,612.10^{16}. \ e^{-0,0225. \ t}$

$\quad -2,69999 = -0,0225. \ t$

$\quad t \approx 120 \text{ sec}$

131. Calculate the activity for the pure mass of 3,5 microgram of the nuclide $^{32}P_{15}$, when the half-life of this nuclide is $1,23.10^6$ sec.

Solution:

$\lambda = 0,693 / \{1,23.10^6 \text{ sec}\}$

$\lambda = 5,63414. \ 10^{-7} \text{ sec}^{-1}$

$N_0 / 3,5.10^{-6} \text{ g} = 6,024.10^{23} / 32$

$N_0 = 6,584375.10^{16}$ nuclei

$(\Delta N/\Delta t) = \lambda N_0$

$\quad\quad = 5,63414. \ 10^{-7} \text{ sec}^{-1}. \ 6,5843.10^{16}$ nuclei

$(\Delta N/\Delta t) = 3,7097332.10^{10}$ decay/sec.

132. The radioactive sample of $^{210}Po_{84}$ isotope is 0,4 milligram. When the half-life of this isotope is 138,376 days. Which mass will be stay from this material after one year?

Solution:

$N = N_0 (1/2)^{t/T1/2}$
$m = m_0 (1/2)^{t/T1/2}$
$\quad = 0,4(1/2)^{365,25/138}$ mg
$\quad = 0,4. \, 0,1596806$
$m= 0,0638722$ mg.

133. How many alpha particles and beta particles which emitted in the series of decay from $^{238}U_{92}$ to $^{206}Pb_{82}$?

Solution:

The relation of the uranium series is:

$A = 4n+ 2$,

The difference between the first and the final mass number is:

$238 - 206 = 32$,

The change in mass number due to alpha-decay is:

$\Delta A = 4$

Hence: $32 / 4= 8$ alpha particles.

And in the same method we can find beta particles:

$92 - 82 = 10$ the difference in atomic number due to beta when:

$\Delta Z = 1$

Therefor: $10/1= 10$ beta particles.

134. How much energy is required to remove a neutron from $^{24}Na_{11}$ nucleus?

Solution:

The atomic mass of the nucleus of $^{24}Na_{11}$ is:

m = 23,990961amu And the daughter nucleus ($^{23}Na_{11}$) has the atomic mass of:
m = 22,989767amu

The mass of a neutron is: m_n = 1,008665amu

Therefore the mass of the final system is:

22,989767amu+ 1,008665amu = 23,990961amu

Hence the mass increase of the system is:

23,998432amu – 23,990961amu = 7,471.10^{-3}amu

Thus the energy required to remove a neutron from the nucleus of $^{24}Na_{11}$ is:

7,471.10^{-3}amu . 931,5Mev/amu = 6,95Mev.

135. If the activity of a radioactive substance is initially 398 dis / min and two days later it is 285 dis / min, what is the activity four days later still, or six days after the start? Give your answer in dis / min.

Solution:

$A = A_0 e^{-\lambda t}$
$285 = 398 e^{-\lambda (2 . 24 . 60)}$
ln (285 / 398) = -λ (2880)
ln (0,7160804) = -λ (2880)
- 0,3339627 = -λ (2880)

Hence:

$\lambda = 1{,}1596 \cdot 10^{-4}$ min^{-1}

We can say that is after 4 days from the two days, when:

$A = A_0\, e^{-\lambda t}$
$\quad = 285\, e^{-\lambda\,(\,1{,}1596.10\text{-}4)(4.\ 24.60)}$
$A = 285 \cdot 0{,}5127713 = 146{,}1398$ dis/min.

Or after 6 days from start:

$A = A_0\, e^{-\lambda t}$
$\quad = 398\, e^{-\,(\,1{,}1596.10\text{-}4)(6.\ 24.60)}$
$\quad = 398\, e^{-1{,}001887}$
$\quad = 398 \cdot 0{,}3671856$
$A = 146{,}13$ dis/min.

136. To make the dial of a watch glow in the dark, 1.10^{-9} kg of radium $^{226}Ra_{88}$ is used. The half-life of this isotope is $1{,}6.10^3$ years. How many kilograms of radium disappear while the watch is in use for fifty years?

Solution:

Activity$= N_0$ / mass $= N_A$ /A
Activity$= N_0$ / $1.10^{-9}.10^3 = 6{,}025.10^{23}$ / 226

Hence:

$N_0 = 1.10^{-6}.6{,}025.10^{23}$ / 226
$N_0 = 2{,}6654867.10^{15}$ nuclei.

When:

$$\lambda = 0,693 / T_{1/2}$$
$$= 0,693 / 1,6.10^3. 365,25.24.3600$$
$$\lambda = 1,37249.10^{-11} \ sec^{-1}$$

When:

$$A_0 = \lambda N_0$$
$$A_0 = (1,37249.10^{-11}) (2,66548.10^{15})$$
$$= 36583,543 \ dis/sec$$
$$A_0 \approx 36584 \ dis/sec.$$
$$A = A_0 \ e^{-\lambda t}$$
$$A = 36583,543. \ e^{- (1,3724902.10-11)(50. \ 365,25.24. \ 3600)}$$
$$= 36584. (0,9785767)$$
$$A \approx 35800 \ dis/sec.$$

When:

$$A = \lambda N$$

Therefor:

$$35800= (1,3725.10^{-11}). N$$

Hence:

$$N = 2,61.10^{15} \ nuclei.$$
$$Activity= N_0 / mass = N_A /A$$
$$Activity= 2,61.10^{15} / mass = 6,025.10^{23} / 226$$

Hence:

$$m = 9,786.10^{-7} gram.$$

137. Calculate the width of the first excited energy level of an isotope, when the energy of this level is 0,87Mev above ground level, and the half-life of this isotope is $1,73.10^{-4}$ μsec.

Solution:

When the mean life time is: $\tau = 1 / \lambda$
And: $\lambda = 0,693 / T_{1/2}$
Hence: $\tau = 1 / \lambda = T_{1/2} / 0,693$
$\tau = 1,73.10^{-4}.10^{-6} / 0,693$
$\tau = 2,5.10^{-10}$ sec

A level width Γ of the excited energy level is defined as:

$\Gamma = (h / 2\pi). \lambda$

Therefore:

$\Gamma = (h / 2\pi \tau)$

Hence:

$\Gamma = \{6,62.10^{-27} / 2. \; 3,14. \; 2,5.10^{-10} \}. \; 1 / 1,6.10^{-12}$
$\Gamma = \{6,62.10^{-27} / 1,57.10^{-9}\}. \; 2,31168.10^{-12}$

Therefore:

$\Gamma = 2,632.10^{-6}$ ev.

138. How many disintegrations occur each second in 0,5 g of a sample of $^{243}Am_{95}$? What is the number of curies in this mass?

Solution:

The half-life of $^{43}Am_{95}$ is 7380 years:

We find the decay constant for this sample:

$\lambda = 0,693 / \{7380. 365,25. 24. 3600\}$
$\quad = 0,693 / 2,329.10^{11}$
$\lambda = 2,9756.10^{-12}$ sec^{-1}

The number of atoms in 0,5 gm of $^{243}Am_{95}$ is:

$N = \{6,025.10^{23}.0,5\} / 243$
$N = 1,2397.10^{21}$ nucleus
Activity$(A) = \lambda N$
$\quad\quad\quad = 2,9756.10^{-12}. 1,2397.10^{21}$
$\quad\quad A = 3,6888.10^{9}$ dis / sec.

We know that each one curie= $3,7.10^{10}$ dis / sec.

Therefore:

Activity $= 3,6888.10^{9} / 3,7.10^{10}$
$\quad\quad A = 0,09969$ curies.

139. Two milligram of Actinium-227($^{227}Ac_{89}$) half-life is 21,773 years, is allowed to decay for 1 year, what is the activity at the end of that time?

Solution:

$A = \lambda N = \lambda N_0 \, e^{-\lambda t}$
$\lambda = 0,693 / 21,773.365,25.24.3600$ sec^{-1}
$\quad = 0,693 / 6,871.10^{8}$
$\lambda = 1,0085.10^{-9}$ sec^{-1}
$t = 1$ year$= 365,25.24.3600$
$t \approx 3,16. 10^{7}$ sec
$N_0 = (6,025.10^{23} / 227). 0,002$
$N_0 = 5,30837.10^{18}$ nuclei
$A = 1,0085.10^{-9}. 5,30837.10^{18}. e^{-(1,0085.10-9). 31557600}$

A = 5,1857822.10⁹ dis/sec
$$A = 5,1857822.10^9 \text{ dis/sec}$$
$$= 5,1857822.10^9 / 3,7.10^{10}$$
$$A = 0,14 \text{ curie}$$

140. How many nucleus of $^{249}Cf_{98}$ isotope that will accumulate in 170 years with a generation rate of 10^{14} nucleus/sec, and how many grams does this comprise?

Solution:

The relation between the number of nucleus and the time of the disintegration of nucleus is:

$dN / dt = g - \lambda N$, when N= 0 , at t= 0

Where g is the constant rate of the generation of atoms per sec, and:

λN: The rate of decay, and the relation is the pure rate of accumulation of particle.

Or:

$g = (dN / dt)+ \lambda N$

When we use the relation of the integration and the constant of this integration , We can write the new form of this relation:

$N_0 = g / \lambda. (1- e^{-\lambda t})$
When $g = 10^{14}$ atoms/sec
$\qquad t = 170.365,25.24.3600$
$\qquad t = 5,364.10^9$ sec.

Hence:

$\lambda = 0,693 / \{351. 365,25. 24.3600\}$
$\qquad = 0,693 / 1,1076.10^{10}$

$\lambda = 6{,}256.10^{-11}$ sec^{-1}
$N_0 = 10^{14} / 6{,}256.10^{-11} (1- e^{-\lambda t})$
$\quad = 1{,}5985.10^{24} (1- e^{-\lambda t})$
$\quad = 1{,}5985.10^{24} (1- e^{-6{,}256.10-11. \, 5{,}364.10^9})$
$N_0 = 1{,}5985.10^{24} (1- e^{-0{,}33557})$
$N_0 = 4{,}5568.10^{23}$

When:

Activity$= N_0 / $ mass $= N_A /A$
And $N_0= \{ N_A /A\}$. Mgrams

Then:

$M = \{N_0. A \} / N_A$
$M = \{4{,}5568.10^{23}. 249 \} / 6{,}025.10^{23}$
$M = 188{,}325$grams.

141. How many disintegrations occur each second in a 0,5 g of a sample of $^{245}Cm_{96}$?

Solution:

The half-life of $^{245}Cm_{96}$ is 8500 year,
We find the decay constant of $^{245}Cm_{96}$

This is:

$\lambda = 0{,}693 / \{8500. 365{,}25. 24.3600\}$
$\quad = 0{,}693 / 2{,}6824.10^{11}$
$\lambda = 2{,}5835.10^{-12}$ sec^{-1}

The number of atoms in 0,5gram of this isotope is:

Activity$= N_0 / $ mass $= N_A /A$
And: $N_0= \{ N_A /A\}$. Mgrams

Then:

$N_0 = \{ 6,025.10^{23} / 249 \}. 0,5$
$N_0 = 1,2296.10^{21}$ nucleus
$\lambda N_0 = 2,5835.10^{-12}. 1,2296.10^{21}$
$\lambda N_0 = 3,1766.10^9$ dis / sec
The activity for each curie is $3,7.10^{10}$ dis/sec

Therefore:

The activity in this case is:

$A = 3,1766.10^9 / 3,7.10^{10}$
$A = 0,085854$ curies.

142. An isotope of ($^{191}Os_{76}$) has a radioactivity of 2,4 curie uses in a purpose for 8,65 days. What is the activity at that time?

Solution:

The decay constant of this isotope is:

$\lambda = 0,693 / T_{1/2}$
$\lambda = 0,693 / \{15,4.24.3600\}$
$\quad = 0,693 / 1330560$
$\lambda = 5,208.10^{-7}$ sec^{-1}

The activity in this case is:

$A_0 = 2,4. 3,7.10^{10}$
$A_0 = 8,88.10^{10}$ dis/sec.
$A = \lambda N = \lambda N_0. e^{-\lambda t}$
$\quad = 8,88.10^{10}. e^{-5,208.10-7(8,65.24.3600)}$
$\quad = 8,88.10^{10}. e^{-(0,389225)}$
$\quad = 8,88.10^{10}. 0,6775819$
$A = 6,0169.10^{10}$ dis/sec

143. The isotope of $^{35}P_{15}$ has a half-life of 14,262 days.
   a) How long the time must take $0,5N_0$ (N_0 the original number of nucleus) to decay?
   b) Find the mean life (τ) of this isotope.

Solution:

We calculate the disintegration constant:

λ = 0,693 / 14,262.24.3600
λ = 0,693 / 1232236,8
λ = 5,624.10^{-7} sec^{-1}

a)   When: $N = N_0\, e^{-\lambda t}$

N= 0,5 N$_0$
$0,5N_0 = N_0\, e^{-\lambda t}$
0,5 $= e^{-\lambda t}$
ln 0,5 = - (5,624.10^{-7})t
t   = -ln 0,5 / 5,624.10^{-7}
         = 1232498,3 sec / 24. 3600
t   = 14,265 days.

b)   The mean life is:

$\tau = 1 / \lambda$
    = 1 / 5,624.10^{-7}
τ = 1778093,8 sec

Therefore:

τ = 20,579 days.

144. What is the fraction of radium emission as a sample of radon, will disintegrate in two days? $^{222}Rn_{86}$ ($T_{1/2}$=3,82 days).

Solution:

$N / N_0 = e^{-\lambda t}$
$\quad = e^{-(0,693 \,.\, 2 \,/\, 3,82)}$
$N / N_0 = e^{-(0,693 \,.\, 2 \,/\, 3,82)}$
$N / N_0 = 0,695$

145. What is the fraction of the original amount remains a radioactive element disintegrates for a period of time equal to its average life?

Solution:

$t = T_{av} = 1/\lambda$

Hence:

$\lambda t = \lambda T_{av} = 1$

And:

$N/N_0 = e^{-\lambda t} = e^{-1} = 1/e$
$N/N_0 \approx 0,37$

146. How much $^{238}U_{92}$ which contains after disintegrating for a period equal to its average life, when a piece of this element is 4 gm.

Solution:

Activity$= N_0 / mass = N_A /A$
Activity$= N_0 / 4 = 6,025.10^{23} / 238$

Hence:

$N_0 = 1,012605.10^{22}$ nuclei
$t = T_{av}$ and $\lambda T_{av} = 1$
$N = N_0 \, e^{-\lambda(Tav)}$
$N = N_0 \, e^{-1}$

Hence:

$N = 1,012605.10^{22}. \, e^{-1}$
$N = 1,012605.10^{22}. \, 0,367879$
$N = 3,725.10^{21}$ nuclei

147. What is the age of the uranium ore if now contains 0,5gm of $^{206}Pb_{82}$ for each gram of $^{238}U_{92}$?

Solution:

As known that the lead is of radioactive origin:

And:

$N = N_0 \, e^{-\lambda(t)}$
$N = 1 - 0,5 = 0,5$ uranium ore.
$N_0 = 1$, $T_{1/2} = 4,468.10^9$ years

By the relation of:

$t = 1 / \lambda_U \ln \{(Npb + Nu)/Nu\}$

And: $\lambda_U = 0,693 / T_{1/2}$
         $\lambda_U = 1,55102.10^{-10}$ year^{-1}

Or:
$\lambda_U = 4,915.10^{-18}$ sec^{-1}

Hence:

$\{(Npb + Nu)/Nu\} = (0,5 + 1)/1 = 1,5$
$t = (1/4,915.10^{-18}) \ln(1,5)$
$= 8,2495422.10^{16}$ sec

Or: $t = 2,614.10^9$ years.

148. What is the mass of one curie of:
   d) $^{235}U_{92}$
   e) $^{210}Po_{84}$
   f) $^{239}Pu_{94}$

   $T_u = 7,038.108 years$
   $T_{po} = 138,376 days$
   $T_{pu} = 24,119 years.$

Solution:

The half-life of $^{235}U_{92}$ is: $7,038.10^8$ years

Hence:

$\lambda_u = 0,693 / 7,038.10^8 . 365,25.24.3600$
   $= 0,693 / 2,2210238.10^{16}$
$\lambda_u = 3,12.10^{-17}$ sec^{-1}
$A_0 = \lambda N_0$
$A_0 = 3,7.10^{10}$ dis/sec $= 3,12.10^{-17}$ sec^{-1}. N_0

Therefore:

$N_0 = 3,7.10^{10} / 3,12.10^{-17}$
$N_0 = 1,1858.10^{27}$ nuclei

When:

Activity$= N_0 / \text{mass} = N_A / A$

Hence:

$m = N_0. A = N_A$
$\quad = 1,1858.10^{27}.235 / 6,025.10^{23}$
$m = 462522,12$ gm

Or:

$m = 462,522$ kg.

The half-life of ^{210}Po is 138,376 days.

$\lambda_{Po} = 0,693 / T_{1/2}$
$\quad = 0,693 / (138,376. 24. 3600)$
$\quad = 0,693 / 11955686$ sec
$\lambda_{Po} = 5,7964.10^{-8}$ sec^{-1}
$A_0 = \lambda N_0$
$3,7.10^{10}$ dis/sec$= 5,7964.10^{-8}. N_0$

Therefore:

$N_0 = 6,38326.10^{17}$ nuclei

When:

$N_0 / m = N_A / A$

Then:

$m = N_0. A / N_A$
$\quad = (6,38326.10^{17}.210) / 6,025.10^{23}$
$m = 2,2248.10^{-4}$ gm

The half-life of ^{239}Pu is 24,119 years

$\lambda_{pu} = 0,693 / T_{1/2}$
$\quad = 0,693 / 24,119.365,25. 24. 3600$
$\quad = 0,693 / 7,61137.10^8$
$\lambda_{pu} = 9,10479.10^{-10} \text{ sec}^{-1}$

And: $A_0 = \lambda N_0$

Therefore:

$3,7.10^{10} = 9,10479.10^{-10} . N_0$

Hence:

$N_0 = 4,06379.10^{19}$ nuclei.

When:

$m = N_0.A / N_A$

Therefore:

$m = 4,06379.10^{19}. 239 / 6,025.10^{23}$
$m = 0,0161202$ gm.

149. a) *Assuming that there are 10 tons of $^{238}U_{92}$ and 15 tons of $^{232}Th_{90}$ uniformly distributed in the first meter of depth under each square kilometre of the surface of the earth. Find the combined activity of these two radioelements in micro curies in the cubic meter of soil under each square meter of surface.*

   b) *Assuming radioactive equilibrium, find the mass of $^{226}Ra_{88}$ in one cubic foot of soil.*

Solution:

a)

The radius of the earth (R)= 6400 km
The area of the earth surface= $4\pi R^2$

$$= 4. 3,14 (6400)^2= 5,14718.10^8 \text{ km}^2$$

The total $^{238}U_{92}$ and ^{232}Th under the first meter of depth of the earth surface is:

$$= 10 \text{ tons. } 5,14718.10^8$$
$$= 5,147.10^9 \text{ tons.}$$
$$= 5,147.10^{15} \text{ gm.}$$

And for Thorium =

$$= 15 \text{ tons. } 5,14718.10^8= 7,72.10^9 \text{ tons.}$$
$$= 7,72.10^{15} \text{ gm.}$$

Now for:

$^{238}U_{92}$: $T_{1/2}= 4,468.10^9$ years

Hence:

$\lambda_U =4,915.10^{-18}$ sec^{-1}

And for ^{232}Th:

$T_{1/2}= 1,405.10^{10}$ years
$\lambda_{Th} = 1,5629.10^{-18}$ sec^{-1}
$N\lambda = 4,915.10^{-18}.(6,025.10^{23} / 238). 5,147.10^{15}$
Activity= $6,4039.10^{19}$ dis/sec.
A $= 1,73. 10^{15}$ micro curies.

The area of the earth surface in square meter is:

$5,147.10^8$ km^2 = $5,147.10^8$ (10^6)
$\qquad\qquad$ = $5,14718.10^{14}$ m^2

The activity of ^{238}U in each cubic meter of soil under each m^2 of the earth surface is from the results:

Activity / m^3= $1,73.10^{15}$ / $5,14718.10^{14}$
$\qquad\qquad$ = 3,3625 µc.

Similarly for the ^{232}Th in each m^3 of soil from the last result by changing the corresponding.

For Thorium:

Activity= λ N
$\qquad$ = $1,5629.10^{-18}.(6,025.10^{23}$ / 232).$7,72077.10^{15}$
$\qquad$ = $3,133.10^{19}$ dis/sec
$\quad$ A = $8,469.10^{14}$ microcurie.
Activity / m^3= $8,469.10^{14}$ / $5,14718.10^{14}$
$\qquad\qquad$ = 1,6453µc

Adding the two activities per m^3 of these two isotopes:

$\qquad\qquad$ = $3,3625863 + 1,6453669 = 5,0079532$µc

b)

$N\lambda$ $\;= N_1\lambda_1 = N_2\lambda_2 = $......etc

$N\lambda$ is given by the number of nucleus of ^{238}U and of ^{226}Ra(which has $T_{1/2}$ = 1600y)
λ_{Ra} $\;= (0,693/5,049216.10^{10})$
$\qquad$ = $1,3725.10^{-11}$ sec^{-1}
$N_1\lambda_1$ = $1,3725.10^{-11}.(6,025.10^{23}/226).$ M$_{Ra}$
$\qquad$ = $1,3725.10^{-11}.(6,025.10^{23}/226).$ M$_{Ra}$
$\qquad$ = $3,65896.10^{10}$ M$_{Ra}$

Therefore:

$M_{Ra} = 4,4968.10^{-11}$ gm.

150. What is the number of alpha decays in one gram sample of Californium $^{249}Cf_{98}$ ($T_{1/2}$= 351years) in 50 years, and in 300 years.

Solution:

When:

$\Delta N = N_0 - N$
$= N_0 - N_0\, e^{-\lambda t}$ (when $N = N_0 e^{-\lambda t}$)
$\Delta N = N_0\,(1 - e^{-\lambda t})$

When:

$\lambda t = (0,693 / T_{1/2})\, t$
$= \{(0,693 / 351.365,25.24.3600)\}.\,(50.365,25.24.3600)$
$\lambda t = 0,0987179$
$\Delta N = N_0\,(1 - e^{-\lambda t})$
$\Delta N = N_0\,(1 - e^{-0,0987179})$
$= N_0\,(1 - 0,9059983)$
$\Delta N = N_0\,(0,0940017)$

We can say as:

$\Delta N \approx N_0 \lambda t$

The number of nuclei in one gram, N_1 is:

$N_1 = (6,025.10^{23} / 249).\,1gram$
$N_1 = 2,4196.\,10^{21}$ nuclei

When the sample of ^{249}Cs considered is 1gram initially $N_0 = N_1$ and:

$\Delta N = 2,4196.10^{21}.(0,0987179)$
$\Delta N = 2,3886.10^{20}$ nuclei.

Therefore:

$2,3886.10^{20} / 2,4196.10^{21} = 0,0987211$
$\lambda t = (0,693 / 351\text{years}). (300 \text{ years})$
$\lambda t = 0,5923076$
$\Delta N = N_0 (1 - e^{-0,5923076})$
$\Delta N = N_0 (0,4469503)$
$\Delta N \approx N_0 \lambda t$
$\Delta N = 2,4196.10^{21}. 0,5923076$
$\Delta N = 1,433.10^{21}$ nuclei

Therefore:

$1,433.10^{21} / 2,4196.10^{21} = 0,59231$

151. Calculate the activity of 5gram of $^{249}Cf_{98}$ and the specific activity of this isotope.

Solution:

Activity $= \lambda N$
$= (0,693 / T_{1/2}). N$
$= (0,693 / 351.365,25.24.3600).\{(6,025.10^{23} / 249). 5\text{gram}\}$
$= (0,693 / 1,1076717.10^{10}). 1,2098.10^{22}$
$= 7,5692.10^{11}$ dis/sec
Activity $= 7,5692.10^{11} / 3,7.10^{10}$
A $= 44,5247$ Ci

The specific activity= activity by curei / mass by gram
$= 44,5247$ Ci / 5gram
$= 8,905$ Ci/g

152. The isotope of $^{60}Co_{27}$ responsible for the activity of beta $2,63.10^8 dis/sec$, which makes up 0,014% of natural mixture. When the mass occurring cobalt is 45 mg naturally. What the half-life of this isotope.

Solution:

m = 45mg
N = (0,014 / 100). 45. 10^{-3}gm. ($6,025.10^{23}$ / A)
  = (0,014 / 100). 45. 10^{-3}gm. ($6,025.10^{23}$ / 60)
N = $6,32625.10^{16}$ nuclei

And the half-life is:

$T_{1/2}$ = 0,693 / λ

When:

dN / dt = λN

And:

1 / λ = N / (dN / dt)

Hence:

$T_{1/2}$ = 0,693. 1/λ
   = 0,693. N / (dN / dt)
   = 0,693. $6,32625.10^{16}$ / {$2,6354.10^8$dis/sec}
   = $1,6635.10^8$ sec
$T_{1/2}$ = 5,27 years.

153. Iodine $^{131}I_{53}$ used in diagnostic and therapeutic techniques in the treatment of thyroid disorders. This isotope has a half-life of 8,04 days. What percentage of an initial sample of this isotope remains after 30 days?

Solution:

$$N = N_0 e^{-\lambda t}$$
$$N / N_0 = e^{-\lambda t}$$

When $T_{1/2} = \ln 2 / \lambda = 0,693 / \lambda$

Then:

8,04 days= $0,693 / \lambda$

Hence:

$\lambda = 0,693 / 8,04$
$\lambda = 0,0862$ days^{-1}

And:

percentage $= N / N_0 = e^{-(0,0862)(30)}$
$N / N_0 = e^{-2,586} = 0,0753$

Therefore:

The percentage is 7,53%

154. Find the separation energy of 2α particles from Radium $^{226}Ra_{88}$?

Solution:

We find first for one α – particle:

$B_\alpha = \{M(A-4,Z-2)+ M(^4He_2) - M(A,Z)\}$ 931,5
$B_\alpha = \{M(222 , 86)+ M(^4He_2) - M(226 , 88)\}$ 931,5
   $= \{(222,017571)+ 4,002602 - 226,025402 \}$ 931,5
   $= \{ - 5,23.10^{-3} \}$ 931,5
$B_\alpha = -4,8717$ Mev.

We need 9,7435Mev in order to separate two alpha particles from Radium, for this negative sign!

155. How many Mev for the separation energy of 1n from $^{24}Na_{11}$?

Solution:

$B_n = \{ M(A-1,Z)+ m_n - M (A,Z)\}$ 931,5
$= \{ M(23 , 11)+ m_n - M (24 , 11)\}$ 931,5
$= \{22,989767+ 1,008665 - 23,990961\}$ 931,5
$= (7,471 \cdot 10^{-3})$. 931,5
$B_n = 6,9592$Mev.
$E_{bind} = a_1A - a_2A^{2/3} - a_3Z^2/A^{1/3} - a_4(A-2Z)^2/A+\delta A^{-3/4}$ (1)

Coefficients in (1) are selected from the conditions of the best agreement of the model curve of the distribution with the experimental data. Since this procedure can be carry ouied differently, there are several collections of the coefficients of Weizsacker's formula. Frequently they are used in (1) as the following

$a_1 = 15.6$ Mev
$a_2 = 17.2$ Mev
$a_3 = 0.72$ Mev
$a_4 = 23.6$ Mev
$\delta = 0$ (A-odd)
$\delta = 34$ Mev(Z - even, N – even)
$\delta = - 34$ Mev(Z - odd, N – odd)

156. Estimate the nuclear binding energy of $^{12}C_6$ according to Weizsacker formula and to compare result with the same value, obtained from the experimental data about the masses.

Solution:

$E_{BE}= a_1A - a_2A^{2/3} - a_3Z^2/A^{1/3} - a_4(A-2Z)^2/A+\delta A^{-3/4}$
$= \mathbf{15,6\ (12) - 17,2\ (12)^{2/3} - \{0,72(6)^2\}\ /\ (12)^{1/3} - [23,6(12-12)^2\ /\ 12\ ]}$
$\mathbf{+34(12)^{-3/4}}$
$= 187,2 - 90,153 - 11,321698+ 5,2734272$
$= 90,998Mev$
$E_{BE} \approx 91Mev.$

From the experimental values of the masses E= 92,2Mev

When the mass of the nucleons in this nucleus is:

$(6\ m_p+ 5m_n)=$
$= \{6(1,007825amu)+ 6(1,008665amu)\}$
$= \{6,04695amu+ 6,05199amu\}$
$= 12,09894amu$

The mass of the nucleus $^{12}C_6$ is 12amu

The defect of mass Δm= 12,09894amu – 12= 0,09894amu

The binding energy of the nucleus is:

$E_{BE} = (\Delta m\ .\ 931,5\ )$
$= (0,09894amu).931,5$
$E_{BE} = 92,16261Mev.$

We can see the difference between these two methods:

92,162 – 90,998= 1,164Mev

This is a few!

157. A radioactive material produces 1280 decay per minute at one time, and 6h later produces 320 decays per minute. What is the half-life of this material?

Solution:

N_0= 1280 dis / min, N= 320 dis / min.
$N = N_0 \, e^{-\lambda t}$
$320 / 1280 = e^{-\lambda t}$
$0,25 = e^{-\lambda t}$
$\ln 0,25 = -\lambda t$
$-1,386294 = - (6.60) . \lambda$

Hence:

$\lambda = 1,386294 / 360$
$\lambda = 3,851.10^{-3} \text{ min}^{-1}$

The half-life is $T_{1/2:}$

$T_{1/2} = 0,693 / \lambda$
$\quad\;\; = 0,693 / 3,851.10^{-3}$
$T_{1/2} = 179,962 \text{ min.}$

158. What is the fraction of a sample of $^{68}Ge_{32}$, whose half-life is about 9 months, will remain after 4,5yr?

Solution:

$N = N_0 \, e^{-\lambda t}$
$N / N_0 = e^{-\lambda t}$
$\quad\quad = e^{-0,693 / 9.30.24.3600} \cdot {}^{4,5. \, 365,25.24.3600}$
$N / N_0 = 67,94084$

Hence the fraction of this sample is:

$N / N_0 = 67,94\ \%$

159. $^{124}Cs_{55}$ **has a half-life of 30,8 sec.**
   a) *If we have 9,5 µg initially, how many nuclei are present?*
   b) *How many are present 2 min later?*
   c) *What is the activity at this time?*
   d) *After how much time wills the activity drop to less than about 1 per second?*

Solution:

a) $\lambda = \ln2 / T_{1/2}$
   $\lambda = 0,693 / 30,8 = 0,0225\ \text{sec}^{-1}$
   $N / \text{mass} = N_A / A$

Then:

$N / 9,5.10^{-6} = 6,025.10^{23} / 124$

Therefore:

$N = 9,5.10^{-6}.\ 6,025.10^{23} / 124$

Then:

$N = 4,616.10^{16}$ nucleus.

b) $N_{2m} = 4,616.10^{16}.\ e^{-0,0225.\ 2.\ 60}$
   $N_{2m} = 4,616.10^{16}.\ 0,0672$

N, which are present 2 min later is:

$N = 3,102.10^{15}$ nuclei.

c) $A = \lambda N$
   $A = 0,0225. \, 4,616.10^{16}$
   $A = 1,0386.10^{15}$ dis/sec
   $A = A_0 \, e^{-\lambda t}$

Hence:

$1 = 1,0386.10^{15}. \, e^{-0,0225 \, t}$

Therefore:

$e^{-0,0225 \, t} = 9,6285.10^{-16}$

And: $-0,0225t = 1$

Hence:

$t = 1 / 0,0225 = 44,44$ sec.

160. The activity of a sample of $^{35}S_{16}$ ($T_{1/2} = 7,56.10^6$ sec) is $6,8.10^6$ decays per second. What is the mass of the sample present?

Solution:

The decay constant of this isotope is:

$\lambda = \ln2 / T_{1/2}$
   $= 0,693 / 7,56.10^6$ sec
$\lambda = 9,1666.10^{-8}$ sec^{-1}

The activity of this isotope is:

$A = \lambda N$

Hence:

$6,8.10^6 = 9,1666.10^{-8}. N$

Then the number of nuclei N is:

$N = 6,8.10^6 / 9,1666.10^{-8}$
$N = 7,418.10^{13}$ nuclei.
$N / mass = N_A / A$

Then:

$7,418.10^{13}$ nuclei / mass $= 6,025.10^{23} / 35$

Therefore:

$m = 7,418.10^{13}.35 / 6,025.10^{23}$

Then:

$m = 4,309.19^{-9}$ gram.

161. A radioactive nuclide produces 2880 decays per minute at one time, and 1,6h later produces 820 decays per minute. What is the half-life of the nuclide?

Solution:

The radioactivity of this isotope is:

$A = A_0 . e^{-\lambda t}$

Hence:

$820 = 2880. e^{-\lambda(1,6.60)}$
$820 / 2880 = e^{-\lambda(96)}$

$0,2847222 = e^{-96\lambda}$

$\ln 0,2847222 = -96\lambda$

Hence:

$\lambda = 0,0131$ min⁻¹

Then:

$T_{1/2} = 0,693 / 0,0131$

$T_{1/2} = 52,9$ min

162. Proton with the kinetic energy E_k of 2Mev strikes stationary nucleus $^{197}Au_{79}$. Determine the differential scattering cross-section $d\sigma/d\Omega$ at an angle $\theta= 60°$ As the value of the differential scattering cross-section will change, if we as the scattering nucleus select $^{27}Al_{13}$?

Solution:

The differential cross section of the elastic Coulomb scattering to the angle is determined by Rutherford's formula:

$d\sigma/d\Omega = [Z_1 Z_2 e^2 / 4 E_k] 1/\sin^4(\theta/2)$

Where:

Z_1 - charge of the collide particle, Z_2 – charge of the nucleus

Then:

$d\sigma/d\Omega = [\{1. 79. 1,44Mev.fm\} / (4. 2Mev)]^2. 1 / (1/2)^4$
≈ 3200 fm² / st
$d\sigma/d\Omega = 326$/st

It follows from Rutherford's formula that the relation of the differential scattering cross-section during the replacement of nucleus ^{197}Au on ^{27}Al will be determined by the relation of the squares of the charges of these nuclei:

$R = (d\sigma/d\Omega)_{Au} / (d\sigma/d\Omega)_{Al}$
$\quad = Z^2_{Au} Z^2_{Al}$
$\quad = (79)^2 / (13)^2$
$R = 37$

163. Calculate the scattering cross section of α - particles with the kinetic energy
$E_k = 5Mev$ **by the coulomb field of** $^{208}Pb_{82}$ **nucleus at angles are more than** 90^0.

Solution:

We will obtain cross section by the integration of Rutherford's formula

$\sigma(\theta>\theta_0) = \int_\Omega (d\sigma/d\Omega)d\Omega$

$$\int_{\theta0}^{\pi} \int_0^{2\pi} (d\sigma/d\Omega)\sin\theta \, d\theta \, d\varphi$$

$$2\pi \int_{\theta0}^{\pi} [Z_1 Z_2 e^2/4 \, E_k]^2 \sin\theta \, d\theta/\sin^4(\theta/2)$$

$= 4\pi[Z_1 Z_2 e^2/4 \, E_k]^2[-1/\sin^2\theta/2]|^{\pi}_{\theta0}$
$= 4\pi[Z_1 Z_2 e^2/4 \, E_k]^2[\{1/\sin^2(\theta_0/2)\} -1$
$4\pi[Z_1 Z_2 e^2/4 \, E_k]^2$
$= 4.3,14((2.82.1,44Mev.fm) / 2. \, 5Mev)^2$
$= 7000fm^2$
$\sigma(\theta>\theta_0) = 706$

164. Calculate the differential cross section $d\sigma/d\Omega$ of an elastic proton scatterings on the nuclei of gold $^{197}Au_{79}$ at the angle of 15°, if it is known that for the section of the irradiation of target with a thickness d= 7 mg/cm² by protons with the summary charge Q=1 nCuloumb to the detector with an area S= 0.5cm², located at a distance of l= 30cm from the target, it fell that ΔN= 1,97 $\cdot 10^5$ of the elastic scattered protons. Compare the experimentally measured section from Ratherford differential cross-section.

Solution:

Form of this reaction is:

a+ A   →   B+ b   is called the value of

$d\sigma_{ab}(\theta,\varphi) / d\Omega = (1 / nN)(dN/d\Omega)$

Where n – the quantity of the particles of the target per unit of area N - a quantity of the fallen on target particles, $dN/d\Omega$ – a quantity of the particles products of this reaction b, which departed into the elementary solid angle $d\Omega$ in the direction, characterized by polar θ and azimuthal φ angles. Differential cross section usually is measured in the barns to the steradian:

$dN/d\Omega=\Delta N/\Delta\Omega$
$\Delta\Omega= S / l^2$
$N= Q/e_p$
$n= d\cdot N_A/A$

Where e_p - charge of a proton, N_A - Avogadro's number and A - the mass number of nucleus ^{197}Au

Differential cross section will be:

$d\sigma_{ab}(\theta,\varphi) / d\Omega=$
$= [Ae_p / dN_A Q][\Delta N\ell^2 / S]$
$= (197g/mol).1,6.10^{-19}C.1,97.10^5.(30cm)^2 / 7.10^{-3}g/cm^2. 6,02.10^{23}. 1/ mol.1.10^{-9}C.0,5cm^2$
$d\sigma_{ab}(\theta,\varphi) / d\Omega = 2.65\cdot10^3$ b/st

The differential cross section of the elastic Coulomb scattering according to Rutherford's formula for the protons with the kinetic energy $E_k = 3 Mev$ is

$d\sigma / d\Omega = [Z_{Au} Z_p e^2 / 4 E_k] (1/\sin^4\theta/2)$
$= \{79.1.1, 44 Mev.fm/4.3 Mev\}^2 [\{1/\sin^4(15^0/2)\}]$
$d\sigma / d\Omega = 3,1. 10^3 b/st$

The obtained value is close to that experimentally measured of the section

165. Determine the cross-section of the reaction of $^{31}P_{15}(n,p)^{31}Si_{14}$, if it is known that after the irradiation target $^{31}P_{15}$ with a thickness of d= 1 g/cm² in the neutron flux J= $2 \cdot 10^{10}$ c⁻¹·cm⁻² in period t_{ir}= 4 h, its β- activity I, measured after the time of t_{ir}= 1 hour after the end of irradiation, proved to be $I(t_{pr})$= $3.9 \cdot 10^6$ decay/s. Period of the half-life $T_{1/2}(^{31}Si_{14})$= 157.3 min.

Solution:

The number of nuclei ^{31}Si, which are formed into 1 s in this reaction, is

$N(t) = \sigma J n = \sigma J (d N_A/A)$

Where n – the number of the nuclei per unit of the area of target

N_A - Avogadro's number
A- mass number of the nuclei of ^{31}Si

The number of those decomposing into 1 from nuclei N(t)

Where:

$\lambda = \ln 2/T_{1/2}$
$= 60.0,693/157,3$
$\lambda = 0.264$ h⁻¹ ; the λ - disintegration constant of ^{31}Si

Then:

$dN(t) / dt = \sigma J (d N_A/A) - \lambda N(t)$

In this case: $N(0) = 0$

We obtain, which up to the moment of the time t_{ir} was formed nuclei ^{31}Si

$N(t_{ir}) = \sigma J (d N_A/\lambda_A) (1 - e^{-\lambda t_{ir}})$

Through the time interval t_{pr} after the end of the irradiation, the number of nuclei ^{31}Si is:

$N(t_{pr}) = N(t_{ir}) e^{-\lambda t_{pr}} = \sigma J (d N_A/\lambda_A) (1 - e^{-\lambda t_{ir}}) e^{-\lambda t_{pr}}$

The activity of the preparation:

$I(t_{pr}) = \sigma J d N_A/A (1 - e^{-\lambda t_{ir}}) e^{-\lambda t_{pr}}$

For the reaction cross-section is obtained:

$\sigma = I(t_{pr}) A / J d N_A (1 - e^{-\lambda t_{ir}}) e^{-\lambda t_{pr}}$
$= 3,9.10^6 \ s^{-1}.31g.mol / 2.10^{10} \ s^{-1}.1g.cm^{-2}. \ 6,02.10^{23}mol^{-1}(1-e^{-0,264h-1.4h})$
$e^{-0,264h-1.1h}$

$\sigma \approx 2 \cdot 10^{-26} \ cm^2 = 20 \ mb$

166. Calculate the thermal fission cross section for an enriched uranium mixture which contains 2% of $^{235}U_{92}$ atoms.

Solution:

We can expresse the fission cross-section for this mixture by the weight average:

$\sigma_f = [N_0(238) \sigma_f(238) + N_0(235) \sigma_f(235)] / [N_0(238) + N_0(235)]$

And the referring to the table of cross section, we get:

$\sigma_f = \sigma_f(235) / [1 + (N_0(238)/N_0(235))]$
   $= 582/(1+98/2)$
$\sigma_f = 11,64$ barns

167. Calculate the fission rate for $^{235}U_{92}$ required to produce 2 watt and the amount of the energy that is released in the complete fission of 1800grams of this isotope.

Solution:

The rate of the required fission is:

$r = 2$ Watt / 200Mev / fission
   $= (2. 10^7$erg / sec) / (200. 1,6. 10^{-6})erg / fission
$r = 6,25. 10^{10}$ fission / sec

The number of nuclei per 1800grams is:

$N = (1800$grams / 235grams). 6,025. 10^{23}

The energy which released is:

$E = (1800 / 235). 6,025.10^{23}. 200$Mev
   $= 9,23. 10^{26}. 1,602. 10^{-13}$ J
Where: 1 Kcal= 4186 J

Hence:

$E = 1,478. 10^{14}$J / 4186
$E = 3,53.10^{10}$ Kcal

168. The Isotope of $^{235}U_{92}$ has a half life of $7,038.10^8$ years for a fission. Estimate the rate of the fission for 1gram of $^{235}U_{92}$.

Solution:

dN/dt = λN = (ln2/T).N

Where N = 6,025. 10^{23}/235 = 2,564. 10^{21}
dN/dt   = 0,693. 2,564. 10^{21} / 7,038.10^8. 3,155.10^7
dN/dt   = 22,22 h^{-1}

169. The cross section for the (n,α) reaction with boron follows the 1/v law. If the cross section for 40 eV neutrons is 16 barns, calculate the cross section for neutrons with 0,024 eV.

Solution:

From this value we can say that these neutrons are slow,

Therefor:

$\sigma = \sigma_0 (E_0/E)^{1/2}$
   = 16(40 /0,024)$^{1/2}$
   = 16. 40,824828
$\sigma \approx$ 653 barns.

170. Calculate the value of the integral section when the differential cross section of the reaction is dσ/dΩ at the angle of 90^0 composes 10 mb/sr

Solution:

If the angular dependence of the differential cross section takes the form:

1+2sin90^0

$$\sigma = \int (d\sigma / d\Omega)\, d\Omega = a \int_0^{2\pi} \int_0^{\pi} (1+2\sin\theta)\sin\theta\, d\theta\, d\varphi$$

$$=2\pi a(\int \sin\theta\, d\theta + 2\int \sin^2\theta\, d\theta)$$
$$= 2\pi a(2+\pi)$$

Let us find constant a from the condition:

$a(1+ 2\sin90°)= 10.$ $a= 10/3$ mb/sr

As a result; we will obtain:

$\sigma = [2.\ \pi.\ 10(2+\pi)] / 3$
$\sigma \approx 108$ mb

171. Calculate the scattering cross section of an α - particle with the energy of 3Mev in the coulomb field of $^{238}U_{92}$ nucleus in the angular interval from 150⁰ to 170⁰

Solution:

We will use Rutherford's formula for the differential elastic cross section of the nonrelativistic charged particle to the angle in the coulomb field of nucleus

$$d\sigma / d\Omega = [Zze^2/4\ E_k]^2.\ 1/\sin^4\theta/2 = C/\sin^4\theta/2$$

Where E_k -the kinetic energy of the impinging particle, z and Z - charges of the impinging particle and target nucleus respectively. The scattering cross section of an α- particle in the angular interval θ_1 -$\theta_2$is:

$\sigma_a = \int(d\sigma / d\Omega)\, d\Omega$
  $= -2\pi C(2/\sin^2\theta/2) \big|_{\theta1}^{\theta2}$
  $= C\int_0^{2\pi} \int_{\theta1}^{\theta2} [1/ \sin^4\theta/2]\sin\theta\, d\theta\, d\varphi$

$=8\pi [Zze^2 / 4T]^2 .[(1/sin^2\theta_2/2) - (1/sin^2\theta_1/2)]$
$=8\pi [92.2. 1,44 / 4. 3]^2 [(1/sin^275^0) - (1/sin^285^0)]$
$\sigma_a= 7,86$ barn

172. The nucleus of $^{10}B_s$ from the excited state with the energy of 0.72Mev decomposed by the emission of γ- quantum with the period of half-life

$T_{1/2}= 6.7 \cdot 10^{-10}s$. **Estimate uncertainty energy ΔE of that emitted γ- quantum**

Solution:

From the uncertainty principle of Heisenberg; we will obtain:

$\Delta E \approx \hbar / \Delta t = \hbar /\tau = \hbar ln2 / T_{1/2}$
$\quad =[6,6.10^{-16}ev.s.0,69]/6,7.10^{-10}$ s
$\Delta E \approx 7. 10^{-7}ev$

Where τ -an average of the life time in this condition

173. Gold plate with a thickness of l= 1 micrometer irradiates by a beam of α- particles with the density of flow j= 10^5 particle/cm^2.s; Kinetic energy of α- particles E_k is 5Mev. How much α- particles per unit of solid angle does fall per second to the detector, which located at the angle θ= 170° to the axis of bundle? Area of the bundle spot on target S is 1cm²

Solution:

The number of the particles which scattered per unit time into the single solid angle equal to:

$N= jSn(d\sigma/d\Omega)$

Where n – the number of nuclei per the unit surface area of the target and$(d\sigma/d\Omega)$- the differential cross section of an elastic scattering

The number of the nuclei per the unit surface area of the target is:

$n = \rho \ell N_A / A$

Where ρ- the density of the target, l - its thickness, A - the mass number of the target materials and N_A - Avogadro number

The beam of the particles through the detector is:

$N = [jS\rho \ell N_A / A][Z_1 Z_2 e^2 / 4 E_k]^2 . 1 / \sin^4(\theta/2)$
   $\{ [10^3 \text{cm}^{-2}.\text{s}^{-1}. 1\text{cm}^2. 19,3\text{g/cm}^3.10^{-4}\text{cm}. 6,02. 10^{23} \text{mol}^{-1}] / 197\text{g/mol}\}$
   $[\{2.79.1,44\text{Mev.fm}. 10^{-13}\text{cm/fm}\}/ \{4.5\text{Mev}\}]^2 . 1,015$
$N \approx 0,77\text{s}^{-1}$

174. Determine the upper boundary of the positrons spectrum, emitted when β^+ - nuclear decomposition of $^{27}Si_{14}$. The energy of β^+ disintegration, by using values of the masses of atoms

Solution:

$Q = M_{at.}(A, Z) - M_{at.}(A, Z - 1) - 2 m_e$

Where $M_{at}(A, Z)$ - the mass of the atom of initial nucleus and $M_{at.}(A, Z - 1)$ - the mass of the atom of nucleus- product (mass in the energy units). The mass of atom of ^{27}Si is equal to 25137.961Mev, and ^{27}Al - 25133.150Mev. Upper boundary of the spectrum of positrons is equal to the energy of the decay

$E_{kmax} = Q$
   $= 25137,961\text{Mev}-25133,150\text{Mev}-2.0,511\text{MeV}$
$E_{kmax} = 3.789 \text{ Mev}$

175. Transfer several nuclear reactions, in which can be formed isotope $^8Be_4$

Solution:

We can get the isotope of ^8Be by the differents reactions, but we must look to the conservations laws of the charge and of the nucleons number; then we obtain:

1. $\alpha + \alpha$ $\rightarrow$ $^8Be + \gamma$, 5. $\gamma + {}^{10}Be$ $\rightarrow$ $^8Be + d$
2. $d + {}^6Li$ $\rightarrow$ $^8Be + \gamma$, 6. $p + {}^{10}B$ $\rightarrow$ $^8Be + {}^3He$
3. $p + {}^7Li$ $\rightarrow$ $^8Be + \gamma$, 7. $p + {}^{11}B$ $\rightarrow$ $^8Be + \alpha$
4. $\gamma + {}^9Be$ $\rightarrow$ $^8Be + n$, 8. $p + {}^{10}B$ $\rightarrow$ $^8Be + \alpha$

176. What is the minimum kinetic energy in the laboratory system $E_{k\,min}$ which must have a neutron so that would become possible reaction $^{16}O_8(n,\alpha)^{13}C_6$?

Solution:

The minimum energy, with which is possible this reaction, is equal to the threshold of the reaction:

Let us calculate the energy of this reaction:

The form of the reaction is:

$$^{16}O + n \rightarrow {}^{13}C + {}^4He_2$$

Then:

$Q = (15,994915amu + 1,008665amu) - (13,003355amu + 4,002602amu)$
  $= 17,00358amu - 17,005957amu$
$Q = -2,377.10^{-3}amu$

Hence:

Q= -2,377.10⁻³· 931,5

Q= -2,214Mev

In order to calculate the threshold energy T_{th} and by the nonrelativistic approximation, we can find T_{th}:

$T_{min} = T_{th} = 2.214\,(1+1/17)$

$T_{min} = 2,214\,(1,0588235)$

$T_{min} \approx 2,345\text{Mev}$

177. Determine the thresholds T_{th} of the reactions of the $^{12}C_6$ photo disintegration

1. $\gamma + {}^{12}C_6 \rightarrow {}^{11}C_6 + n$
2. $\gamma + {}^{12}C_6 \rightarrow {}^{11}B_5 + p$
3. $\gamma + {}^{14}C_6 \rightarrow {}^{12}C_6 + n + n$

Solution:

Let us calculate the reaction energies (1 – 3)

By using the data in the table on the collide masses of atoms:

1. $Q = \{(0+12)-(11,011433+1,008665)\}.931,5$
   $= 931,5\,\{12-12,020098\}$
   $= (-0,020098).931,5$
   $Q = -18,721\text{Mev}$

2. $Q = \{(0+12)-(11,009305+1,007825)\}.931,5$
   $= \{12-12,01713\}.931,5$
   $= (-0,01713).931,5$
   $Q = -15,956595\text{Mev}$

3. Q= {(0+ 14,003242) - (12+ 2(1,008665)}. 931,5
    = {14,003247-14,01733}.931,5
    = (0,014088 -).931,5
    Q= -13,122972Mev

For the threshold energy and by using the relations, we can say that is possible to write down:

$T_{th} \approx Q$

As for the reactions 1 - 3

$m_a = m_\gamma = 0$

And hence
$<<2m_Ac^2\ Q\ ||$

178. Determine the thresholds of the two reactions: $^7Li_3(p,\alpha)^4He_2$ and $^7Li_3(p,\gamma)^8Be_4$

Solution:

Let us calculate the energy of the reactions as the same method:

$^7Li(p,\alpha)^4He$

Q = {(7,016928+ 1,007825) - (4,002602+ 4,002602}. 931,5
    = {8,024753-8,005204}.931,5
    =(0,019549).931,5
Q =+ 18,209893Mev
$^7Li(p,\gamma)^8Be$

Q = {(7,016928+ 1,007825) - (0+ 8,014555}. 931,5
    = {8,024753-8,014555}.931,5
    = {0,010198}.931,5
Q =+ 9,499437Mev

The two reactions are exothermic with any energy of protons

179. What is the minimum energy which a proton must have; that would become possible the following reaction: $p+d \rightarrow p+p+n$

Solution:

The energy of the reaction is:

$Q = \Lambda\,(^1H) + \Lambda\,(^2H) - 2\Lambda(^1H) - \Lambda(n)$
$Q = \{(1,007825 + 2,014102) - (2(1,007825) + 1,008665\}.\,931,5$
$\quad = \{3,021927 - 3,024315\}.931,5$
$\quad = \{-2,388.10^{-3}\ u\}.931,5$
$Q = -2,2244\,Mev$

Since:

By using the expression, we will obtain:

$<<2m_p c^2\ Q\ ||$
$E_{min} = T_{th} = 2.2244\ (1 + 0.5) = 3.3366\ Mev$

180. Is the reactions are possible
$\qquad \alpha + \,^7Li_3 \quad \rightarrow \qquad ^{10}B_5 + n$
$\qquad \alpha + \,^{12}C_6 \quad \rightarrow \qquad ^{14}N_7 + d$

Solution:

Under the action α- particles with the kinetic energy $E_k = 7\,Mev$

Thresholds of the reactions:

$\alpha + \,^7Li \rightarrow \quad ^{10}B + n$
$Q = \{(4,002602 + 7,016003) - (10,012936 + 1,008665\}.\,931,5$
$\quad = \{11,018605 - 11,021601\}.931,5$
$\quad = \{-2,996.10^{-3}\ u\}.931,5$
$Q = -2,790774\,Mev$
$E_{th} = 2.791\ (1 + 4/7) = 4.386\ Mev$

Reaction is possible, since:

$E_k = 7$ Mev $> E_{th}$

$\alpha + {}^{12}C \quad \rightarrow \quad {}^{14}N + d$

$Q = \{(4,002602 + 12) - (14,003074 + 2,014102\}. 931,5$
   $= \{16,002602 - 16,017176\}.931,5$
   $= (-0,014574 \ u).931,5$

$Q = -13,575681$ Mev

$E_{th} = 13,576 \ (1 + 4/7) = 18,1$ Mev

$E_k = 7$ Mev $< E_{th}$

$E_{th} = 18.1$ Mev

Reaction is impossible, since:

$E_k < E_{th}$

181. Calculate the energy Q of the following reactions and, identify the particle of X

Solution:

a) ${}^{12}C_6 + X_1 \quad \rightarrow \quad {}^{13}N_7 + \gamma$

b) ${}^{15}N_7 + {}^{1}H_1 \quad \rightarrow \quad {}^{12}C_6 + X_2$

c) ${}^{19}Ne_{10} \quad \rightarrow \quad {}^{19}F_9 + X_3 + \nu$

d) ${}^{35}Cl + X_4 \quad \rightarrow \quad {}^{32}S + \alpha$

In order to identify the particle of X, it is necessary to use laws of the charge conservation and the number of nucleons

a) ${}^{12}C_6 + {}^{1}H_1 \quad \rightarrow \quad {}^{13}N_7 + \gamma$

$A_1 = (13 + 0) - 12 = 1$

$Z_1 = (7 + 0) - 6 = 1$

New form of the reaction is

a) $^{12}C_6 + {}^1H_1 \rightarrow {}^{13}N_7 + \gamma$

$Q = [12 + (1,007825) - (13,005738) + 0].931,5$
$= \{13,007825 - 13,005738\}.931,5$
$= 2,087.10^{-3}.931,5$
$Q = 1,944 Mev$

Nucleus	Atomic number Z	Atomic mass A	Q:The reaction's energy,Mev
$^{12}C_6$	6	12	
X_1	1	1	
$^{13}N_7$	7	13	
γ	0	0	
			+1,944 exothermic reaction

b) $^{15}N_7 + {}^1H_1 \rightarrow {}^{12}C_6 + X_2$

$A_2 = (15+1) - 12 = 4$
$Z_2 = (7+1) - 6 = 2$

The form of the reaction is:

b) $^{15}N_7 + {}^1H_1 \rightarrow {}^{12}C_6 + \alpha({}^4He_2)$

$Q = [15,000108 + (1,007825) - \{(12) + 4,002602\}].931,5$
$= \{16,007933 - 16,002602\}.931,5$
$= 5,331.10^{-3}.931,5$
$Q = 4,9658265 Mev$

Nucleus	Atomic number Z	Atomic mass A	Q:The reaction's energy,Mev
$^{15}N_7$	7	15	
$^1H_1$	1	1	
$^{12}C_6$	6	12	
X_2	2	4	
			+4,966 exothermic reaction

c) $^{19}Ne_{10} \quad \rightarrow \quad ^{19}F_9 + e^+ + \nu$

$A_3 = (19 - 19) = 0$
$Z_3 = (10 - 9) = 1$

The new form of the reaction is:

$^{19}Ne_{10} \quad \rightarrow \quad ^{19}F_9 + e^+ + \nu$
$Q = [19,001879 - (18,998404) - \{5,485931.10\text{-}4\} + 0\}]. \, 931,5$
$\quad = \{2,9264.10^{-3}\}.931,5$
$Q = 2,726Mev$

Nucleus	Atomic number Z	Atomic mass A	Q:The reaction's energy,Mev
$^{19}Ne_{10}$	10	19	
$^{19}F_9$	9	19	
X_3	1	0	
ν	0	0	
			+2,726 exothermic reaction

d) $^{35}Cl + X_4 \quad \rightarrow \quad ^{32}S + \alpha$

$A_4 = (32 + 4) - 35 = 1$
$Z_4 = (16 + 2) - 17 = 1$

The form of the reaction is:

$$^{35}Cl_{17} + {}^1H_1 \rightarrow {}^{32}S_{16+} \alpha$$
$$Q = [34,968853 + (1,007825) - \{(31,972071) + 4,002602\}]. 931,5$$
$$= \{35,976678 - 35,974673\}.931,5$$
$$= 2,005.10^{-3}.931,5$$
$$Q = 1,8676575 Mev$$

Nucleus	Atomic number Z	Atomic mass A	Q:The reaction's energy,Mev
$^{35}Cl_{17}$	17	35	
$^1H_1$	1	1	
$^{32}S_{16}$	16	32	
α	2	4	
			+1,8677 exothermic reaction

182. Calculate the threshold of the following reaction

$$^{14}N_7 + \alpha \rightarrow {}^{17}O_8 + P \qquad \textit{in two cases}$$

If the collide particle is:

An α- particle

Nucleus $^{14}N_7$, and the energy of the reaction is Q= 1.15 Mev

Solution:

Let us calculate the threshold energy:
$$E_{kth} = 1.15(1 + 4/14)$$
$$= 1.478 \text{ Mev}$$
$$E_{kth} = 1.15(1 + 14/4)$$
$$E_{kth} = 5.175 \text{ Mev}$$

In the first case on the motion of the center of inertia «uselessly» is expended (4/14)Q, in the second (14/4)Q, thus the threshold of the reaction in the second case is higher 3.5 times

183. Find the types of the energies and thresholds of the following reaction

$$^{32}S_{16}(\gamma,a)^{28}Si_{14}$$
$$^{7}Li_{3}(p,n)^{7}Be_{4}$$
$$^{4}He_{2}(a,p)^{7}Li_{3}$$

Solution:

$^{32}S(\gamma,a)^{28}Si$

We can write these reactions as:

a) $^{32}S_{16}+\gamma \rightarrow {}^{28}Si_{14}+\alpha$

The new form of the reaction is:

$Q = [31,972071+ (0) - (27,976927+ 4,002602)]. 931,5$
  $= \{31,972071-31,979529\}.931,5$
  $= -7,458.10^{-3}.931,5$
$Q = - 6,947Mev$

Nucleus	Atomic number Z	Atomic mass A	Q:The reaction's energy,Mev
$^{32}S_{16}$	16	32	
γ	0	0	
$^{28}Si_{14}$	14	28	
α	2	4	
			- 6,947 endothermic reaction

^{7}Li(p,n)^{7}Be

We can write these reactions as:

a) $^7Li_3 + {}^1H_1 \rightarrow {}^7Be_4 + {}^1n_0$

The new form of the reaction is:

$Q = [7{,}016003 + (1{,}007825) - (7{,}016928 + 1{,}008665)] \cdot 931{,}5$
$= \{8{,}023828 - 8{,}025593\} \cdot 931{,}5$
$= -1{,}765 \cdot 10^{-3} \cdot 931{,}5$
$Q = -1{,}6441 \, Mev$

Nucleus	Atomic number Z	Atomic mass A	Q:The reaction's energy,Mev
$^7Li_3$	3	7	
$^1H_1$	1	1	
$^7Be_4$	4	7	
$^1n_0$	0	1	-1,6441 endothermic reaction

^{4}He(α,p)^{7}Li
We can write these reactions as:

a) $^4He_2 + {}^4He_2 \rightarrow {}^7Li_3 + {}^1H_1$

The new form of the reaction is:

$Q = [2 (4{,}002602) - (7{,}016003 + 1{,}007825)] \cdot 931{,}5$
$= \{8{,}005204 - 8{,}023828\} \cdot 931{,}5$
$= -0{,}018624 \cdot 931{,}5$
$Q = -17{,}3483 \, Mev$

187

Nucleus	Atomic number Z	Atomic mass A	Q:The reaction's energy,Mev
$^4He_2$	2	4	
$^7Li_3$	3	7	
$^1H_1$	1	1	
$^1n_0$	0	1	
			-17,3483 endothermic reaction

184. What is the energy of γ- quant when a slow neutron collide the $^7Li_3$ nucleus?

Solution:

The reaction is:

$^7Li(n,\gamma)^8Li$

The form of this reaction is:

$^7Li_3 + {}^1n_0 \rightarrow {}^8Li_3 + \gamma$

By the same method to find the energy of this reaction:

$Q = [7,016003 + (1,008665) - (8.0224855 + 0)]. 931,5$
$= \{8,024668 - 8,0224855\}.931,5$
$= 2,1825.10^{-3}.931,5$
$Q = 2,033Mev$

When:

$P_\gamma = E_\gamma / c$, $P_N = (2M_N E_N)^{1/2}$ and $E_N + E_\gamma = Q$

Hence:

$$|P_N| \approx |P_\gamma|$$

Where:

P_N ,P_γ the pulses of the nucleus and of γ-quant, and E_N , E_γ energy of the nucleus and of γ-quant, then

$$E_N = E^2_\gamma / 2M_N c^2$$
$$\approx Q^2/2M_N c^2$$
$$\approx (2,033)^2/(2.7.931,5)$$
$$E_N \approx 3,169. \ 10^{-4} Mev$$

The energy of γ- quant is:

$$E_\gamma = Q - E_{Li}$$
$$E_\gamma \approx Q = 2.033 \ Mev$$

When:

$$E_N \approx 0$$

185. Find the kinetic energy of $^9Be_4$ nucleus in the laboratory system which is formed with the threshold value of the neutron energy in the reaction of $^{12}C(n,\alpha)^9Be_4$

Solution:

If the energy of the impact particle equal to the threshold energy, and the energy of the particle, products and respectively their momenta into central system equal to zero.

In the laboratory system momentum of ^9Be ; P'_{Be} equal to the momentum of the translational motion $P^{\backslash p.t}_{Be}$

$P^{\backslash}_{Be} = P^{\backslash p.t}_{Be} + {}^{\backslash}P^{\backslash}_{Be} = P^{\backslash p.t}_{Be}$

$P^{\backslash p.t}_{Be}$ The pulse of the translational motion

Let us express through the momentum of the neutron

$P^{\backslash}_{Be} = \{ m_{Be} / m_{\alpha} + m_{Be} \} P^{\backslash}_{n}$

Let us express in the last expression, the momenta through the energies, and then instead of the neutron energy will substitute expression for the threshold energy:

$\sqrt{(2m_{Be}E_{kBe})} = [m_{Be} / (m_{\alpha} + m_{Be})]\sqrt{(2m_{n}E_{kn})}$
$= \{m_{Be} / (m_{\alpha} + m_{Be})\}\sqrt{(2m_{n}| Q|\{m_{c} + m_{n} /m_{c}\})}$
$Q = [12 + (1,008665) - (9,012182 + 4,002602)]. 931,5$
$= \{13,008665 - 13,014784\}.931,5$
$= -6,119.10^{-3}.931,5$
$Q = - 5,6998 Mev$

We will obtain taking into account that:

$m_{c} + m_{n} \approx m_{Be} + m_{\alpha}$

Finally we will obtain

$E_{kBe} = [m_{n}m_{\alpha} / \{m_{c}(m_{\alpha} + m_{Be})\}] |Q|$
$= [1.4/\{12.(4+9)\}] |-5,6998|$
$E_{kBe} = - 0,146 Mev$

Inverse of alpha particle:

186. The sign of this result is negativ, that means the direction of the nucleus is inverse of alpha particle

Which the radioactive isotope were formed in these nuclear reactions?

$^{11}B(t,2p)$?, $^{11}B(t,2n)$?, $^{11}B(a,t)$?

Solution :

$^{11}B(t,2p)$?

$A_1 = (11+3) - 2 = 12$
$Z_1 = (5+1) - 2 = 4$

The radioactive is $_4eB^{12}$

The new form of this reaction is:

$^{11}B(t,2p)^{12}Be_4$

The second reaction is:

$^{11}B(t,2n)$?

$A_2 = (11 + 3) - 2 = 12$
$Z_2 = (5 + 1) - 0 = 6$

The radioactive isotope is $_6C^{12}$

The new form of this reaction is:

$^{11}B(t,2n)^{12}C_6$

The third reaction is:

$^{11}B(\alpha,t)$?

$A_3 = (11 + 4) - 3 = 12$
$Z_3 = (5 + 2) - 1 = 6$

The radioactive isotope is: $_6C^{12}$

The new form of this reaction is:

$^{11}B(\alpha,t)\ ^{12}C_6$

187. Alpha particle with the kinetic energy E_k of 20Mev experiences elastic head-on collision with the nucleus of $^{12}C_6$ Determine the kinetic energy in hp of nucleus $^{12}C_6$ E_{kC} after collision

Solution:

We will use the formula about the general reaction: a+ A $\rightarrow$ b+ B

$E_{kb} = \{m_a m_b E_{ka} / (m_b + m_B)^2\}$
$\{\cos\theta_b \pm (\sqrt{\cos\theta_b^2 + [(m_b + m_B)(m_B - m_a)E_{ka} + m_B Q] / m_a m_b E_{ka})]\}^2$

When the form of the reaction is:

a+ A $\rightarrow$ b+ B

Here $a \equiv \alpha$ and $A \equiv {}^{12}C$

And for the elastic scattering $m_\alpha = m_b$, $m_c = m_B$, $Q = 0$

We will obtain:

$E_{kC} = [4K. E_{k\alpha} / (1+K)^2]. \cos^2 \theta_C$

Where:

$k = m_\alpha / m_C$
θ_C - the angle of emission of nucleus ^{12}C

From the condition of the quation:

$\theta_c = 180^0$

Finally we have:

$E_{kc} = 4.(4/12).20 / (1+(4/12))^2$
$E_{kc} = 15Mev$

188. Determine the maximum and the minimum energy of nuclei of $^7Be_4$ that is formed in the reaction of $^{14}N(n,p)^{14}C$ when $Q= 1,65$ Mev under the action of the accelerated protons with the energy of $E_{kn}= 6$ Mev

Solution:

We will use the formula of:

$E_{kb}= \{m_a m_b E_{ka} / (m_b+ m_B)^2\}$
$\{cos\theta_b \pm (\sqrt{cos\theta_b^2+ [(m_b+m_B)(m_B-m_a)E_{ka}+m_BQ] / m_a m_b E_{ka})]\}^2$
$E_{kC}= m_n m_C E_{kn} / (m_C+m_p)^2 [cos\theta_C \pm \{\sqrt{cos\theta^2_C+[\{(m_C+m_p)\{(m_p-m_n)}$
$E_{kn}+m_pQ\}\}/m_n m_C E_{kn}]]^2$

If the second term under the root negatively, then the angular range is limited to the condition of nonnegative expression under the root:

$cos^2\theta^{max}_C = |(m_p+m_C)[(m_p-m_n) E_{kn}+m_pQ]/m_n m_C E_{kn}|$
$cos^2\theta^{max}_C=|(1+14).[(1-1).6+1.(-1,65)]/1. 14. 6|$
$cos^2\theta^{max}_C= 0,2946$

Hence:

$cos\theta^{max}_C= 0,5428$, $\theta= 57,124$

The maximum angle, at which will depart the nuclei of ^{14}C is $cos\theta^{max}_C$ ~57^0. The kinetic energy of the nuclei of ^{14}C which depart at this angle is:

$E_{kC}= m_n m_C E_{kn.} cos^2\theta^{max}_C / (m_p+m_C)^2$
     $= 1.14.6.0,2946/(1-14)^2$
$E_{kC} \approx 0,11 Mev$

189. The energy of alpha particles is $E_{ka}= 10Mev$, which interact with stationary nucleus of $^7Li_3$. Determine the values of momentum in C.M. system when the form of the reaction is $^7Li(\alpha,n)^{10}B$

Solution:

The form of the reaction is:

$$^7Li_3 + \alpha \rightarrow {}^{10}B_5 + n$$

We calculate the energy Q of this reaction:

$$Q = [7,016003 + (4,002602) - (10,012936 + 1,008665)]. 931,5$$
$$= \{11,018605 - 11,021601\}.931,5$$
$$= -2,996.10^{-3}.931,5$$
$$Q = -2,790774 Mev$$

In order to find E_{kn} and E_{kB} in C.M. system, we must use the formula of:

$$E_{kb}^{\backslash} = M_B /(M_B + M_b) [\{M_A E_{k\alpha} /(M_a + M_A)\} + Q] = [M_B / M_b] E_B^{\backslash}$$

When: $a + A \rightarrow b + B$

And: $a \equiv \alpha$, $A \equiv {}^7Li_3$, $b \equiv n$, $B \equiv {}^{10}B$

Hence:

$$P = (2EM)^{1/2}$$
$$P^2 = 2EM$$
And: $m_n = 1$

Therefore:

$$E_{kn}^{\backslash} = M_B /(M_n + M_B)[\{M_{Li} E_{k\alpha} /(M_\alpha + M_{Li})\} + Q] = (P_n^{\backslash})^2 / 2M_n$$

Hence:

$$P_B^/ = P_n^/ = \sqrt{(2M_n T_n^/)}$$
$$= 1/c[\sqrt{\{(2M_n c^2)(M_B / M_n + M_B)[(M_{Li} T_\alpha /M_\alpha + M_{Li}) + Q]\}}$$
$$\approx 1/c \sqrt{\{(2. 939,566)(10/1 + 10). [7.10 / (4+7) - 2,790774]\}}$$
$$\approx 1/c \sqrt{\{(1879,132)(0,9090909). [3,5728623]\}}$$
$$\approx 1/c \sqrt{\{(1879,132)(3,2480566]\}}$$
$$P_B^/ = 78,125 Mev / c$$

190. Determine the energy of the protons which observed in the reaction of $^{32}S(\alpha,p)^{35}Cl$ at angles of o0 and 90^0 with E= 8Mev. The excited states $^{35}Cl_{17}$ (1,219; 1,763; 2,646; 2,694; 3,003; 3,163Mev) The situation will be excited on the beam of α- particles with energy of 5Mev?

Solution:

The form of this reaction is:

$$^{32}S_{16} + {}^{4}He_2 \rightarrow {}^{35}Cl_{17} + {}^{1}H_1$$

The energy of the reaction Q is:

$Q = (\Delta m_\alpha + \Delta m_s - \Delta m_p - \Delta m_{Cl}) c^2$
$\Delta m_\alpha = 4,002602 - 4 = 0,002602$amu
$\Delta m_s = 31,972071 - 32 = -0,027929$amu
$\Delta m_p = 1,007825 - 1 = 0,007825$amu
$\Delta m_{Cl} = 34,968853 - 35 = -0,031147$amu
$Q = (0,002602 - 0,027929 - 0,007825 + 0,031147). 931,5$
$Q = -2,005.10\text{-}3 . 931,5$
$Q = -1,86766$Mev

The kinetic energy of the colliding particles in C.M. system is:

$E_k^{\backslash} = \{ms / m_\alpha + m_s\} E_{k\alpha}$
$E_k^{\backslash} = \{32 / 4 + 32\} 5,2$
$E_k^{\backslash} = 4,622$Mev

The maximum excitation energy of the nucleus is:

$E_{ex}^{max} = E_k^{\backslash} - |Q|$
$\qquad = 4,622 - 1,868$
$E_{ex}^{max} = 2,754$Mev

With this value of the energy of the colliding particles (5,2)Mev can be excited only states with 1,22 * 1,76 * 2,65 * 2,7Mev. But the energy of the protons, which scattered at angles of 0^0 and 90^0 in this reaction, is calculated by:

$E_{kp}(0^0)= m_p m_\alpha E_{k\alpha} / (m_p+m_{cl})^2 [1+ \sqrt{1+\{(m_p+m_{cl})(Qm_{cl}+(m_{cl}-m_\alpha))} E_{k\alpha} /m_p m_\alpha E_{k\alpha}\}]^2$

$=1.4.5,2/(1+35)^2.1+\sqrt{1+\{(1+35)((-1,868-1,22).35+(35-4).5,2/1.4.5,2)\}}]^2$
$=\{1.4.5,2/(1+35)^2\}\sqrt{91,999}$
$=0,0160493 .(10,5916)^2$
$E_{kp}(0^0)= 1,8Mev$
$E_{kp}(90^0)= [Qm_{cl}+(m_{cl}-m_\alpha) E_{k\alpha}] / (m_p+m_{cl})$
$=[(-3,087).35+(35-4).5,2]/(1+35)$
$=\{-108,045+(31.5,2)\}/36)$
$E_{kp}(90^0)= 1,476Mev$

Where the energy Q of the reaction is:

$Q_1 = Q_0 - E_{ex}$
$Q_1 = -1,868 – 1,22= -3,088$ Mev

And:

$Q_2= -1,868 -1,76= -3,628$ Mev

Analogously for the other cases we can find other values

191.The kinetic energy of a proton is $E_{kp}= 7Mev$ collide the nucleus of $^1H_1$ and the proton scattered with an elastic scattering which the kinetic energy E_k of the B nucleus and scattering angle θ_B of the recoil nucleus $^1H_1$ if the scattering angle of $^1H_1$ is $\theta_b= 40^0$

Solution:

We write the reaction as:

$a + A \rightarrow b + B$

We get for an elastic scattering:

$$E_{ka} = E_{kb} + E_{Kb}$$

Where E_{ka}, E_{kb}, and E_{kB} - kinetic energies of colliding proton, the scattered proton and the nucleus of hydrogen after scattering in L.S. system. From the relation of this reaction; we have:

$$E_{kb} = \{m_a m_b E_{ka} / (m_b + m_B)^2\}\{\cos\theta_b \pm (\sqrt{\cos\theta_b^2 + [(m_b + m_B)(m_B - m_a)}$$
$$E_{ka} + m_B Q] / m_a m_b E_{ka})]\}^2$$
$$E_{kb} = E_{ka} \cos^2\theta_b$$
$$E_{kB} = E_{ka} \cos^2\theta_B$$

As a result we will obtain:

$$E_{kB} = E_{ka} - E_{ka} \cos^2\theta_b$$
$$E_{kB} = E_{ka(} 1 - \cos^2\theta_{b)}$$

When:

$$\sin^2\theta + \cos^2\theta = 1 \text{ , and } \sin^2\theta = 1 - \cos^2\theta$$

Therefore:

$$E_{kB} = E_{ka} \sin^2\theta_b$$
$$= 7\sin^2 40^0$$
$$= 7(0{,}64278)^2$$
$$E_{kB} = 2{,}892 \text{Mev.}$$
$$\theta_B = \arccos(\sin\theta_b)$$
$$= \arccos(\sin 40^0)$$
$$\theta_B = 50^0$$

192. Determine the energy of a neutron E_{kn} which scattered at the angle of 90^0 in the neutron generator and uses the deutrons which accelerated to the energy of $E_{kd} = 0{,}5 Mev$, when the reaction is t(d,n)α.

Solution:

The form of this reaction is:

$$^{3}T_{1} + {}^{2}D_{1} \quad \longrightarrow \quad {}^{4}He_{2} + {}^{1}n_{0}$$

We will determine the energy of this reaction Q:

$Q = \{(3,016049 + 2,014102) - (4,002602 + 1,008665)\}. 931,5$
$Q = \{(5,030151) - (5,011267)\}. 931,5$
$Q = \{0,018884\}. 931,5$
$Q = 17,59 Mev$

Let us determine the energy of: this reaction:

It is utilized relationship of the kinetic energy for the neutron, we will obtain:

$E_{kn}(90^{0}) = \{(m_{\alpha} - m_{d})E_{kd} + m_{\alpha}Q\} / (m_{\alpha} + m_{n)}$
$\qquad = [(17,59. 4 + (4-2). 0,5] / (4+1)$
$\qquad = 71,36 / 5$
$E_{kn}(90^{0}) = 14,272 Mev.$

193. How much the energy of two deutrons which will be released in a reaction. Combine to form an alpha- particle?

Solution:

The energy which released is $= (2. 2,014102) - (4,002602)$
When the mass of $^{2}D_{1}$ is 2,014102amu and of an alpha is 4,002602amu

Hence:

$Q = 4,028204 - 4,002602$
$Q = 0,025602 amu$
$Q = 0,025602. 931,5$
$Q = 23,85 Mev.$

194. Calculate the power which will result from a fission rate of 3. 10^{12} per second.

Solution:

We know that each fission produce about 200Mev of energy:

$= 200. \ 10^6$ ev
$= 2. \ 10^8 . \ 1,6. \ 10^{-12}$erg

Thus, the power released is:

$= 2. \ 10^8. \ 1,6. \ 10^{-12} . \ 3. \ 10^{12}$erg/sec
$= 2. \ 10^8. \ 1,6. \ 10^{-12} . \ 3. \ 10^{12}. \ 10^{-7}$
$W \approx 96$ watt

195. Calculate ζ for $^9Be_4$ – moderator.

Solution:

$\zeta = 1+\{(A-1)^2 \ / \ 2A\} \ \ln \{(A-1)/(A+1)\}$
$= 1+\{(9-1)^2 \ / \ 2. \ 9\} \ \ln \{(9-1)/(9+1)\}$
$= 1+ (64/18) \ \ln (8/10)$
$= 1+ 3,55 . \ (- \ 0,2232)$
$\zeta \approx 0,21$

196.   How much $^{235}U_{92}$ has been consumed in a reactor which has operated for 8 years at an average power of 180 watts.

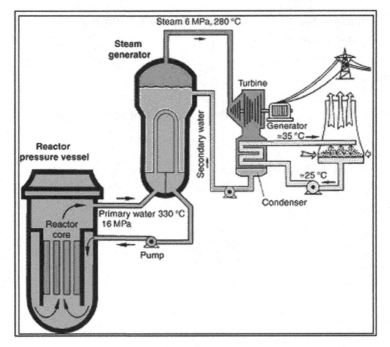

http://blogs.princeton.edu/chm333/f2006/
nuclear/04_nuclear_reactors/01_overview_of_reactors/

Solution:

The total energy which produce= 180. 365,25. 8
$$= 525960 \text{ watt-day}$$

Since 1gram of ^{235}U is equivalent to 1 Megawatt per day;

Therefore:

^{235}U which consumed= 525960 / 10^6
$$= 0,526 \text{ gm.}$$

197. How many fissions per second are required to produce one watt of power?

Solution:

Since 1 fission generates 200Mev of energy 1 watt=
$$= 1 \text{ joule/sec} = 10^7 \text{erg/sec}$$
$$= 10^7 / 1{,}6. \ 10^{-6} = 6{,}25. \ 10^{12} \text{Mev/sec}$$
$$= 6{,}25. \ 10^{12}/200 = 3{,}1.10^{10} \ \text{fission/sec.watt}$$

198. How much power is generated by the fissioning of onegram of $^{235}U_{92}$ per 8 hour.

Solution:

One watt will fission $(3{,}1. \ 10^{10})(8)(3600)=$

$$= 8{,}928. \ 10^{14} \text{ atoms of } ^{235}U \text{ per this time:}$$

In onegram of ^{235}U we have:

Number of the atoms= $6{,}03. \ 10^{23}/235 = 2{,}57. \ 10^{21}$ atom
The energy release $2{,}57.10^{21}/ \ 8{,}928.10^{14} \approx 2878584$ watts .

199. Calculate the number of the collisions required to reduce the fast fission neutrons with an average initial energy of 2Mev to an energy $E_t = 0{,}024$ ev in Berillium (as a moderated assembly for Be_4 is $\zeta = 0{,}209$).

Solution:

$$n= \ln(2. \ 10^6/0{,}024)/0{,}209$$
$$n = 18{,}23836 / 0{,}209$$
$$n \approx 87$$

200. What would be the energy of the neutrons that have 45 collisions with Be_4 nuclei which start with an initial energy of 2Mev?

Solution:

Since n= ln $(E_0 - \ln E'_t) / \zeta$
$$= (1/ \zeta)\ln (E_0 / E'_t)$$
$$E'_t = E_0 \ e^{-n\zeta}$$
$$= 2. \ 10^6 . \ e^{-45. \ 0,209}$$
$$= 2. \ 10^6 .8,23. \ 10^{-5}$$
$$E'_t \approx 308 \ ev$$

201. What is the maximum fractional energy loss for the neutrons in the colissions with $^{39}K_{19}$ nuclei?

Solution:

The mass of ^{39}K nuclei is 39;

Since:

$$(\Delta E / E_0)_{max} = 1-(A-1)^2 / (A+1)^2$$
$$= 1-(39-1)^2 / (39+1)^2$$
$$(\Delta E / E_0)_{max} = 0,097 \approx 9\%$$

202. Calculate the average number of the fission neutrons per neutron absorbed in a uranium mixture which contains $^{235}U_{92}$ and $^{238}U_{92}$ isotopes in 1:12 ratio.

Solution:

The regeneration factor (η) is given by:

$$\eta = \upsilon \ (\sigma_f / \sigma_a)$$

Where υ is the number of neutrons released per fission which is for ^{235}U is equal to 2,43.

Since:

$\sigma_a = [N_0(235) \sigma_a(235) + N_0(238) \sigma_a(238)] / [N_0(235) + N_0(238)]$
$\sigma_a = [\sigma_a(235) + (N_0(238)/N_0(235) \sigma_a(238)] / [1 + (N_0(238)/N_0(235)]$
$\sigma_a = 683 + 12.2,73 / 1 + 12$
$\sigma_a \approx 55,1$ barns

And:

$\sigma_f = \sigma_f(235) / [1 + (N_0(238)/N_0(235)]$
     $= 582/13 = 44,77$ barns
$\eta = 2,43 . (44,77 / 55,1)$
$\eta = 1,974$
$\eta \approx 2$

203. What is the ratio of the released energy from the nuclear fission of $^{235}U_{92}$ to its rest energy , when the released energy in this process is about 200Mev.

Solution:

The number of protons in $^{235}U_{92}$ nucleus is:

$Zm_p = 92(1,007825)$amu
$Zm_p = 92,7199$amu

And the number of neutrons in this nucleus is:

$Nm_n = 143(1,008665)$amu
$Nm_n = 144,23909$amu

Hence:

$(Zm_p + Nm_n) =$
$$= 92,7199 + 144,23909$$
$(Zm_p + Nm_n) = 236,95899 amu$

And:

$\Delta E = 236,95899 - 235,043924$
$\Delta E = 1,91507 amu$

The rest energy of the nucleus is:

$1,91507. \ 931,5 = 1783,8877 Mev.$

And the ratio is:

$r = 200 / 1783,8877 = 0,1121.$
$r = 11,21\%.$

204. What is the combined mass of the two fragments and there energy in the nuclear fission process to the $^{235}U_{92}$ nucleus when the released energy during this process is about 225Mev.

Solution:

The nuclear process as the following form:

$^{1}n_0 + {}^{235}U_{92} \rightarrow ({}^{A1}X_{Z1} + {}^{A2}Y_{Z2}) + 3({}^{1}n_0) + Q$

The atomic weightof the released material is:

$^{1}n_0 + {}^{235}U_{92} = 1,008665 + 235,043924$
$$= 236,05258 \ amu.$$

And the atomic weight of the produced materials is:

$$(^{A1}X_{Z1} + {}^{A2}Y_{Z2})_{\text{at.w.}} + 3(1{,}008665)\text{amu} + [225/931{,}5]$$

Where:

$Q = (225/931{,}5)$amu is the released energy

$Q = 0{,}2415458$amu

$M_{(\text{comb. Of X+Y})} = 236{,}05258 - [3{,}025995 + 0{,}2415458]$
$\qquad\qquad\quad = 236{,}05258 - 3{,}2675408$
$\qquad\qquad\quad = 232{,}78505$amu The combined mass of the fragments.

The energy of these two fragments is:

$Q_{(X+Y)} = 232{,}78505 \cdot 931{,}5$
$Q_{(X+Y)} = 216839{,}26$Mev.

205. The nucleus $^{235}U_{92}$ absorbs a thermal neutron and make a nuclear fission process into Rubidium $^{93}Rb_{37}$ and Cesium $^{141}Cs_{55}$. What are the nucleons produces by this process and how many are there?

Solution:

$^{235}U_{92} + {}^{1}n_0 \rightarrow {}^{93}Rb_{37} + {}^{141}Cs_{55} + {}^{A}X_Z$
$^{A}X_Z: \quad A = (235+1) - (93+141) = 2$
$\qquad\quad Z = (92+0) - (37+55) = 0$
Then they are two neutrons $[1(^{1}n_0)]$

Hence:

The process will take this form:

$^{235}U_{92} + {}^{1}n_0 \rightarrow {}^{93}Rb_{37} + {}^{141}Cs_{55} + 2({}^{1}n_0)$

206. Determine the number of the neutrons which released during the following fission reaction:

$$^{235}U_{92} + {}^1n_0 \rightarrow {}^{133}Sb_{51} + {}^{99}Nb_{41} + ?({}^1n_0)$$

Solution:

When they are neutrons, so that means:

$A = 1 , Z = 0$

And the number of neutrons is:

$A = (1 + 235) - (133 + 99) = 4$

They are four.

$$^{235}U_{92} + {}^1n_0 \rightarrow {}^{133}Sb_{51} + {}^{99}Nb_{41} + 4({}^1n_0)$$

207.   Consider the induced nuclear reaction:

$$^2H_1 + {}^{14}N_7 \rightarrow {}^{12}C_6 + {}^4He_2$$

And determine the energy in Mev which released when $^{12}C_6$ and $^4He_2$ nuclei are formed in this manner.

Solution:

$$^2H_1 + {}^{14}N_7 \rightarrow {}^{12}C_6 + {}^4He_2 + Q$$

Hence:

$$m_{(2H1+ 14N7)} = (2,014102 + 14,003074)$$
$$= 16,017176 \text{ amu}$$
$$\text{And } m_{(12C6+ 4He2)} = (12 + 4,002603)$$
$$= 16,002603 \text{amu}$$

Hence:

$$\Delta m = 16,017176 \text{ amu} - 16,002603 \text{ u}$$
$$= 0,014573 \text{ amu}$$
$$Q = (\Delta m) . 931,5 \quad \text{Mev.}$$
$$= 0,014573 . 931,5$$
$$Q = 13,578 \text{Mev.}$$

208. During a nuclear reaction, an unknown particle is absorbed by a copper $^{63}Cu_{29}$ nucleus, and the reaction products are $^{62}Cu_{29}$, a neutron and a proton. What is the name, the atomic number and the nucleon number of the compound nucleus?

Solution:

$$^{63}Cu_{29} + {}^{A}X_{Z} \rightarrow {}^{62}Cu_{29} + {}^{1}H_{1} + {}^{1}n_{0}$$
$${}^{A}X_{Z}: \quad A = (62 + 1 + 1) - 63 = 1$$
$$Z = (29 + 0 + 0) - 29 = 1$$

Hence :

$${}^{A}X_{Z} \text{ is } {}^{1}H_{1}$$

And the compound nucleus is:

Zinc: $^{64}Zn_{30}$.

209. Write the reactions below in the shorthand form:

$$^{1}n_{0} + {}^{14}N_{7} \rightarrow {}^{14}C_{6} + {}^{1}H_{1}$$
$$^{1}n_{0} + {}^{238}U_{92} \rightarrow {}^{239}U_{92} + \gamma$$
$$^{1}n_{0} + {}^{24}Mg_{12} \rightarrow {}^{23}Na_{11} + {}^{2}H_{1}$$

Solution:

$$^{14}N_7(n,p)\,^{14}C_6$$
$$^{238}U_{92}(n,\gamma)\,^{239}U_{92}$$
$$^{24}Mg_{12}(n,D)\,^{23}Na_{11}$$

210. Complete the following nuclear reactions, assuming that the unknown quantity signified by the question mark is a single entity:

$$^{43}Ca_{24}\,(\alpha,?)\,^{46}Sc_{21}$$
$$^{9}Be_4(?,n)\,^{12}C_6$$
$$^{9}Be_4\,(p,\alpha)\,?$$
$$?\,(\alpha,p)\,^{17}O_8$$
$$^{55}Mn_{25}(n,\gamma)\,?$$

Solution:

$$^{43}Ca_{20}+\,^{4}He_2 \rightarrow\,^{46}Sc_{21}+\,^{1}H_1$$

Hence:

$$^{43}Ca_{20}\,(\alpha,p)\,^{46}Sc_{21}$$
$$^{9}Be_4 + ? \rightarrow\,^{12}C_6 +\,^{1}n_0$$
$$Z\,(?)\,(6+0)-4=2$$
$$A\,(?)\,(12+1)-9=4$$

Hence:

? is $^{4}He_2 : \alpha$

And we can say that the form of this reaction become as:

$$^{9}Be_4\,(\alpha,n)\,^{12}C_6$$
$$^{9}Be_4 +\,^{1}H_1 \rightarrow\,^{4}He_2 + ?$$
$$Z\,(?)=(4+1)-2=3$$
$$A\,(?)=(9+1)-4=6$$

Hence:

? is $^6X_3 = {}^6Li_3$

And we can say that the form of this reaction become as:

$^9Be_4 (p , \alpha) {}^6Li_3$

$? + {}^4He_2 \rightarrow {}^{17}O_8 + {}^1H_1$
$Z(?) = (8+ 1) - 2 = 7$
$A(?) = (17+ 1) - 4 = 14$
$? = {}^{14}X_7 = {}^{14}N_7$

Hence:

$^{14}N_7 (\alpha , P) {}^{17}O_8$
$^{55}Mn_{25} + {}^1n_0 \rightarrow ? + \gamma$
$Z(?) = (25+ 0) - 0 = 25$
$A(?) = (55+1) - 0 = 56$
$?$ is $^{56}X_{25} = {}^{56}Mn_{25}$

Hence:

$^{55}Mn_{25} (n,\gamma) {}^{56}Mn_{25}$

211. What is the nucleon number A in the reaction of:

$$^{27}Al_{13} (\alpha , n) {}^AP_{15} ?$$

Solution:

The equation of reaction is:

$^{27}Al_{13} + {}^4He_2 \rightarrow {}^AP_{15} + {}^1n_0$

Hence:

$A = (27 + 4) - 1 = 30$
$Z = (13 + 2) - 15 = 0$

Hence:

$^A P_{15}$ is $^{30}P_{15}$.

212. What is the atomic number Z and the element X in the reaction $^{10}B_5(\alpha,p) \ ^{13}X_Z$?

Solution:

This reaction is:

$^{10}B_5 + {^4He_2} \rightarrow {^{13}X_Z} + {^1H_1}$

And:

$Z = (5 + 2) - 1 = 6$

Hence $^{13}X_6$ is Carbon

The $^{13}X_Z$ nuclied is $^{13}C_6$

213. In the following reaction:

$$^1n_0 + {^{10}B_5} \rightarrow {^7Li_3} + {^4He_2}$$

Produce $^7Li_3$ nucleus with an alpha particle. The kinetic energy of $^1n_0$ is slow, and the $^{10}B_5$ nucleus in the ground state. The kinetic energy of alpha particle is $9,3.10^6$ m/s. Calculate

> a) **The kinetic energy of $^7Li_3$ nucleus and**
> b) **The produced energy of this reaction.**

Solution:

From the appendix:

$M_\alpha = 4,00264$amu; $E_{kn} = 0$; $M_n = 1,0087$amu ; $M_{Li} = 7,0160$amu

a) When the neutron and Boron is in the ground state, hence the total momentum befor the reaction is zero, and in this case:

$M_{Li}\, v_{Li} = M_{He}\, v_{He}$
$E_{KLi} = 1/2 M_{Li}\, v^2_{Li}$
$\quad = \frac{1}{2} M_{Li}\, (M_{He}\, v_{He} / M_{Li})^2$
$E_{Kli} = M^2_{He}\, v^2_{He} / 2M_{Li}$

By the substitution and by change the unit of the mass from (amu) to (kg) when:

$1,6.10^{-13}$ J= 1Mev.
$E_{KLi} = \{(4,00264)^2(1,66.10^{-27}kg/amu)^2(9,3.\ 10^6m/s)^2\ \} / \{2(7,0160amu)$
$\quad\quad (1,66.10^{27}kg/amu)\}$
$E_{KLi} = 1,64.10^{-13}$J= 1,02Mev

a) When:

$\quad\quad E_{k\alpha} = E_{kx} = 0$

Hence:

$\quad\quad Q = E_{KLi} + E_{KHe}$

And when:

$E_{KHe} = \frac{1}{2} M_{He}\, v^2_{He} = \frac{1}{2}(4,0026amu)(1,66.10^{-27}kg/amu)(9,3.10^6m/s)^2$
$E_{KHe} = 2,84.10^{-13}$J= 1,78Mev.
$\quad Q = 1,02Mev + 1,78Mev$
$\quad Q = 2,8Mev.$

214. In the nuclear medicine, the maximum permissible concentration of the radium emission for the continuous exposure in the environment is

10^7 micro curies per milliliter of the air. What is the radon content of milliliter for this concentration of the standard air.

Solution:

We will find the mass and the volume to 10^{-8} microcurie of this isotope in the standard conditions, and $\lambda = 2,1.10^{-6}$.

Now:

$10^{-8}\mu ci = 10^{-8}.\ 10^{-6} = 10^{-14}$ curie.
And 1 curie = $3,7.10^{10}$ dis/sec

Hence:

$10^{-8}\mu Ci = 10^{-14}.\ 3,7.10^{10} = 3,7.10^{-4}$ dis/sec.

And:

$3,7.\ 10^{-4} = \lambda n = 2,1.10^{-6}.\ (6,025.10^{23} / 222).\ M(gm)$
And $m = (3,7.10^{-4}.\ 222 / 2,1.\ 10^{-6}.\ 6,025.\ 10^{23})$
$\qquad = (3,7.222 / 2,1.\ 6,025).\ 10^{-21}$ gm.
$m = 65.\ 10^{-21}$ gram of radon.........(1)

The volume of this mass in (1) in the standards condition is:

$(65.10^{-21} / 222).\ 22400 \approx 65.10^{-19}$ cm^3.........(2)

In $1,2929.10^{-3}$ gram of the air is $6,5.10^{-22}$ of radon.

The percentage is: $5.10^{-15}\%$.

When we change 1 liter to cubic cemtimeter:

1liter= 1000cm³
1 ml= 1cm³

In one milliliter of the air is $6,5.10^{-20}$ of radon or $6,5.10^{-16}$ %

215. Is the energy enough to occur this reaction?

$$^{13}C_6 + {}^1H_1 \rightarrow {}^{13}N_7 + {}^1n_0$$

Solution:

$$^{13}C_6 + {}^1H_1 \rightarrow {}^{13}N_7 + {}^1n_0 + Q(?)$$

When the $^{13}C_6$ nucleus is bombarded by 2,2Mev protons.

And from the tabel of isotopes we can take the masses of all:

13,003355amu+ 1,007825amu = 14,011180amu

And to the productors:

13,005738amu+ 1,008665amu= 14,014403amu

Hence:

14,011180amu - 14,014403amu = $-3,223.10^{-3}$amu
-3,223. 10^{-3}amu . 931,5Mev /amu = -3,0022Mev.

Hence:

Q= -3,0022Mev.

When the sign is negative, that means the energy of protons (2,2Mev) is not enough to make this reaction.

216. Calculate the energy which released in the following reaction.

$$^{14}C_7 \rightarrow {}^{14}N_7 + \beta$$

Solution:

To the neutral atom we must conserve the calculation of the electrons; here these are 6 electrons which are around the nucleus. Therefor the mass of the carbon is 14,003242amu and to the daughter nucleus in this decay is 14,003074amu

Hence:

14,003242amu – 14,003074amu= 1,68. 10^{-4} amu

And:

The energy is:

$E = 1,68.10^{-4} . 931,5$
$E = 0,156492$Mev.

217. The nucleus of $^{32}P_{15}$ decay according to the following reaction, what is the daughter nucleus and its atomic mass withamu units, when the kinetic energy of the electron is maximum and it is of 1,71Mev?

$$^{32}P_{15} \rightarrow X_{16} + \beta$$

Solution:

For the $^{32}P_{15}$:

The number of the protons in the $^{32}P_{15}$ nuclied is:

$15m_p = 15(1,007825$amu$) = 15,117375$amu

And the number of the nutrons in this nuclied is:

$17m_n = 17.(1,008665amu) = 17,147305amu$

Hence the number of nucleons in this nucleus is:

$15m_p + 17m_n = 15,117375amu + 17,147305amu$
$$= 32,26468amu$$
$32,26468 . 931,5 = 30054,549Mev$
$30054,549Mev - 1,71 = 30052,839Mev$
$30052,839Mev / 931,5Mev/amu = 32,262843amu$

Therefor this nucleus is sulfur $(^{32}S_{16})$ and the form of this reaction is:

$^{32}P_{15} \rightarrow \,^{32}S_{16} + \beta^-$

218. What is the released energy in the following reaction , when the nucleus decay with an alpha particle or with a beta particle. The mass of $^{218}Po_{84}$ nucleus is 218,008965amu

$^{218}Po_{84} \rightarrow \,^{214}X_{82} + \,^4He_2(\alpha) + Q$
$^{218}Po_{84} \rightarrow \,^{218}X_{85} + \beta + Q$

Solution:

1: For an alpha particle $(^4He_2)$:

$2m_p = 2(1,007825amu) = 2,01565amu$
$2m_n = 2(1,008665amu) = 2,01733amu$
$M_{4He2} => 2m_p + 2m_n$
$$= 2,01565amu + 2,01733amu$$
$M_{4He2} = 4,03298amu$
$\Delta m = 218,008965amu - 213,999798amu - 4,06596amu$
$\Delta m = -0,05679amu$

Therefor:

E_{BE} = -0,05679amu . 931,5Mev/amu
E_{BE} = -52,899885Mev.

This is the energy which we need to make this reaction.

2:   About another reaction:

Δm= 218,008965amu – 218,00868amu – m_{e^-}
Δm= 218,008965amu – 218,00868amu – 5,4859317.10^{-4}amu
Δm= -2,6859317.10^{-4}amu

Therefor:

E_{BE} = Δm . 931,5Mev/amu
E_{BE} = -2,6859317.10^{-4}amu . 931,5Mev/amu
E_{BE} = -0,2501945Mev.

This is the energy which we need to make this reaction.

219. Calculate the released energy in an electron capture process by Berellium nucleus.

Solution:

The form of the process is:

$^{7}Be_{4}$+ $^{0}e^-$ → $^{7}Li_{3}$+ v

The mass of $^{7}Be_{4}$ nucleus is:

M_{7Be4}= 7,016928amu
m_{0e^-} = 5,4859317. 10^{-4}amu
M_{7Li3}= 7,016003amu

Hence:

$Q = (M_{7Be4} + m_{0e-} - M_{7Li3})c^2$
$= (7,016928amu + 5,4859317.10^{-4}amu - 7,016003amu)c^2$
$= (1,4735.10^{-3})c^2 . 931,5Mev/amu$
$Q = 1,3725652Mev.$

220. How many collisions with a carbon nucleus are required to reduce the energy of a fast neutron from 2Mev to 0,04 eV?

Solution:

As we know that for values of A>10 , an approximate expression for Y is:

$Y = 2 / \{A+(2/3)\} = 2 / \{12+(2/3)\}$
$Y = 0,158$

The total reduction in the energy= $2. 10^6 / 0,04 = 5. 10^7$
The average number of the collisions= $\ln(5. 10^7) / Y$
$= 17,72 / 0,158$
The average number of the collisions ≈ 112.

221. How many collisions must make before neutrons speed reduced below the average speed of a hydrogen molecule at 20°C when these neutrons moves with an initial speed of $10^9 cm/sec$ through the hydrogen gas and one of them loses exactly 1/4 of its speed in the each collision.

Solution:

$E_f = E_i (V_f^2 / V_i^2)^n , \quad V_f = (3/4) V_i$
$(V_f^2 / V_i^2) = (3/4)^2$
$E_f / E_i = (3/4)^n = (22.10^4)2/(10^9)2$
$E_f / E_i = (22)^2/10^{10}$

Hence:

$$\log(22)^2 - \log 10^{10} = n(\log 4 - \log 9)$$
$$2(1,3424) - (10) = n\{0,6021 - 0,9542\}$$

And:

$$n = 7,3152 / 0,3521$$
$$n \approx 21 \text{ collisions.}$$

222. Complete the following reactions:

i) $^{211}Pb_{82}$ $\rightarrow$ $^{211}Bi_{83} + ?$
j) $^{11}C_{6}$ $\rightarrow$ $^{11}B_{5} + ?$
k) $^{231}Th_{90}*$ $\rightarrow$ $^{231}Th_{90} + ?$
l) $^{210}Po_{84}$ $\rightarrow$ $^{206}Pb_{82} + ?$
m) $^{2}D_{1} + ^{2}D_{1}$ $\rightarrow$ $? + ^{1}n_{0} + Q$
n) $^{9}Be_{4} + ^{4}He_{2}$ $\rightarrow$ $? + ^{1}n_{0}$
o) $^{31}P_{15} + ^{1}n_{0}$ $\rightarrow$ $^{32}P_{15} \rightarrow ? + ^{0}e_{-1}$
p) $^{24}Na_{11}$ $\rightarrow$ $^{24}Na_{12} + ?$

Solution:

a) $^{211}Pb_{82}$ $\rightarrow$ $^{211}Bi_{83} + \beta^-$
b) $^{11}C_{6}$ $\rightarrow$ $^{11}B_{5} + \beta^+$
c) $^{231}Th_{90}*$ $\rightarrow$ $^{231}Th_{90} + \gamma$
d) $^{210}Po_{84}$ $\rightarrow$ $^{206}Pb_{82} + \alpha(^4He_2)$
e) $^{2}D_{1} + ^{2}D_{1}$ $\rightarrow$ $^{3}He_{2} + ^{1}n_{0} + Q$
f) $^{9}Be_{4} + ^{4}He_{2}$ $\rightarrow$ $^{12}C_{6} + ^{1}n_{0}$
g) $^{31}P_{15} + ^{1}n_{0}$ $\rightarrow$ $^{32}P_{15} \rightarrow ^{32}S_{16} + ^{0}e_{-1}$
h) $^{24}Na_{11}$ $\rightarrow$ $^{24}Na_{12} + ^{0}e_{-1}$

218

223. In the nuclear fusion operation two nucleus of a deterium make the following reaction:

$$^2H_I + {}^2H_I \rightarrow {}^3He_2 + {}^1n_0$$

Calculate the valeu of the energy which release in this operation.

Solution:

From the table of physical constants the mass of this nuclied is: 2,0141024amu

Therefor tha mass of two nuclieds before the reaction is:

m_a = 2. 2,0141024amu
m_a = 4,028204amu

And after the reaction, the total mass is:

m_b = 3,01603amu+ 1,008665amu
m_b = 4,024695amu

The mass defect is:

$\Delta m = m_a - m_b$
      = 4,028204amu - 4,024695amu
Δm = 0,003509amu

Therefor:

The released energy in this reaction is:

$\Delta E_0 = \Delta m. c^2$
      $= 0,003509. 1,66.10^{-27} . (3.10^8)^2$ J
$\Delta E_0 = 5,24.10^{-13}$J
      $\approx 0,524$ PJ
$\Delta E_0 \approx 3,27$Mev.

224. Two deutrons ($^2H_1$ nucleus) combine to form an alpha-particle. Which the energy will be released in this reaction?

Solution:

The form of this reaction is:

$$^2H_1 + {}^2H_1 \rightarrow {}^4He_2$$

The atomic mass of $^2H_1 = 2,014102$amu
The atomic mass of $^4He_2 = 4,002602$amu
The energy which released in this process(E)=
   = {(2. 2,014102amu - 4,002602amu)}. 931,5Mev/amu
   =(4,028204 - 4,002602amu). 931,5Mev/amu= 23,848263Mev.
E ≈ 24Mev.

225. Find the Q-value of the following nuclear reaction:

$$^{231}Pa_{91} \rightarrow {}^{227}Ac_{89} + {}^4He_2 + Q$$

When the energy of an alpha particle is 6,2Mev.

Solution:

By the law of the conservation of the momentum:

$$P_\alpha + P_N = 0,$$

When P_N is the momentum of the recoil nucleus($^{227}Ac_{89}$).

And by the law of the conservation of the energy:

$$E_\alpha + E_N = Q$$

When:

$(P_N)^2 = 2E_N.m_N$

Then:

$E_N = (P_N)^2 / 2m_N$

And $E_N = \{(P_\alpha)^2 / 2m_\alpha\}. \{m_\alpha / m_N\}$

Therefore:

$Q = E_\alpha + \{ E_\alpha (m_\alpha / m_N)\}$
$\quad = E_\alpha \{ 1 + (m_\alpha / m_N)\}$
$\quad = 6,2Mev \{ 1 + (4 / 227)\}$
$\quad = 6,2. 1,0176211$
$Q \approx 6,31Mev.$

226. What is the energy of gamma rays which accompanies the decay of $^{40}K_{19}$ as the following:

$$^{40}K_{19} \rightarrow {}^{0}e_{-1} + {}^{40}Ca_{20}$$

Solution:

The atomic mass of $^{40}K_{19}$ is:

$39,964amu \quad {}^{0}e_{-1} = 5,4859317.10^{-4}amu$

And:

$M_{Ca} = 39,962591amu$
$\Delta m = 39,964amu - (5,4859.10^{-4}amu + 39,9626amu)$
$\quad = 39,964amu - 39,963139amu =$
$\quad = 8,61.10^{-4}amu$
$\Delta m = 802 \text{ kev}.$

This energy is an energy of gamma rays which emits from $^{40}Ca_{20}$ to $^{40}Ca_{20}$ and which it is very few. We can say that the reaction process is stable as:

$$^{40}K_{19} \rightarrow \beta^- + {}^{40}Ca_{20} + \gamma$$

227. Calculate the neutron velocity that corresponds to the energy of 4Mev in the the fission process of $^{239}Pu_{94}$ neucleus.

Solution:

The energy of 4Mev= 4. $1,602.10^{-6}$erg

Therefore:

$E_n = 1/2m\ v^2$
$4.1,602.10^{-6} = 1/2(1,6749286.10^{-24}g)\ v^2$

Hence:

$v = \sqrt{\{6,408.10^{-6} / 8,374643.10^{-25}\}}$
$v = \sqrt{\{7,65166.10^{18}\}}$
$v = 2,766.10^9$cm/sec.

228. A device used in a radiation therapy for cancer contains 0,5 g of cobalt $^{60}Co_{27}$. The half-life of this isotope is 5,27years. Determine the activity of the radioactive material.

Solution:

$N_0 / m =$ Avogadro's number $/ A$
Hence:
$N_0 = m.N_A / A$
$= 0,5.\ 6,025.10^{23} / 60$

And:

$N_0 = 5,02.10^{21}$ nuclei.

When:

$\lambda = 0,693 / T_{1/2}$
$\quad = 0,693 / 5,27. 365.25. 24.3600$
$\quad = 0,693 / 1,663.10^8$
$\lambda = 4,16695.10^{-9}$ sec^{-1}

And:

$A = \lambda N$
$A_0 = \lambda N_0$
$\quad = 4,16695.10^{-9}$ sec^{-1}. $5,02.10^{21}$ nuclei.
$A_0 = 2,0918.10^{13}$ dis/sec.

229. Find the Q- value of the nuclear reactions as:

$^{226}Ra_{88} \rightarrow {}^{222}Rn_{86} + {}^4He_2 + Q$, *when the energy of an alpha particle is 6,2Mev.*

Solution:

By the law of the conservation of the momentum:

$P_\alpha + P_N = 0$

When P_N the momentum of recoil nucleus($^{234}Th_{90}$):

And the law of the conservation of the energy:

$E_\alpha + E_N = Q$

Dr. Mouaiyad M.S.Alabed

When:

$$(P_N)^2 = 2 E_N. m_N$$

Hence:

$$E_N = (P_N)^2 / 2m_N$$
$$E_N = \{(P_\alpha)^2 / 2m_\alpha\}. \{ m_\alpha / m_N\},$$

Hence:

$$Q = E_\alpha + E_\alpha \{ m_\alpha / m_N\}$$
$$= E_\alpha \{ 1 + (m_\alpha / m_N)\}$$
$$= 6{,}2 Mev\{ 1 + (4 / 227)\}$$
$$= 6{,}2.\ 1{,}01762$$
$$Q \approx 6{,}3092 Mev.$$

230. In the fission of $^{235}U_{92}$ nucleus with two fragements are 4 neutrons and 2 beta particles. Find the fragements when these are equal.

Solution:

The reaction form is:

$$^{235}U_{92} + {}^1n_0 \rightarrow {}^{A1}X_{z1} + {}^{A2}Y_{z2} + 4{}^1n_0 + 2{}^0e_{-1}$$

The mass number of productors is:

$$A_1 + A_2 + 4 + 0 = 236$$

Therefore:

$$A_1 + A_2 = 232$$

And the atomic number of productors is:

$Z_1 + Z_2 + 4(0) + 2(-1) = 92$
$Z_1 + Z_2 - 2 = 92$
$Z_1 + Z_2 = 94$

We can see the yield graph and get the pair fragements which they have:

$A_1 + A_2 = 232$

And:

$Z_1 + Z_2 = 94$

The form of this process is:

$^{84}Kr_{36} + {}^{148}Ce_{58} + 4n + {}^{0}e_{-1} + Q$

231. Show that the kinetic energy of a neutron in the nuclear reaction of alpha particle and Berelium is 5,7Mev, when the product nucleus in this reaction is $^{12}C_6$.

Solution:

The nuclear reaction is:

$^{4}He_2 + {}^{9}Be_4 \rightarrow {}^{12}C_6 + {}^{1}n_0$

The atomic masses of $^{4}He_2$ and $^{9}Be_4$ are 4,002602amu , 9,012182amu respectively.

Hence:

The atomic mass of them is:

4,002602amu + 9,012182amu = 13,014784amu

And in the same state for the productors:

$12 + 1,008665 = 13,008665$amu

Therefore the difference of the mass is:

$\Delta m = 13,014784$amu $- 13,008665$amu $= 6,119.10^{-3}$amu

But 1amu $= 931,5$Mev.

Hence:

The kinetic energy is:

$E_k = 6,119.10^{-3}$amu . $931,5$Mev/amu
$\approx 6,599$Mev.
$E_k \approx 6,6$Mev

232. A neutron with an energy of 4Mev forward to the graphite and collide with it. How many collisions are required for this particle to loose on the average of 95% of an initial energy?

Solution:

The number of the collisions is:

$n = (1 / \zeta) \ln (100 / 5)$
$= (1 / \zeta) \ln (20)$

Where ζ for the carbon is 0,158.

Hence:

$\zeta = (1 / 0,158). 2,9957$
$\zeta = 18,96$

233. How many collisions with a carbon nucleus are required to reduce the energy of a fast neutron from 2Mev to 1/30 ev?

Solution:

For A >10 an approximate expression for ζ is:

$\zeta = 2 / \{A+(2/3)\}$
$= 2 /\{12+(2/3)\}$
$\zeta = 0{,}158$

The total reduction in the energy=

$= 2. 10^6 / (1/30)$
$= 2.10^6 / 0{,}033333$
$= 6.10^7$ ev

The average number of collisions=

$= \ln (\text{total reduction in the energy}) / \zeta$
$= \ln (6.10^7) / 0{,}158$
$= 17{,}909 / 0{,}158$
n $\approx 113{,}35$

234. What the angle of θ_n relative to the direction of a proton beam will collide neutrons with this energy. When the neutron production are used the reaction of $^7Li_3(p,n)^7Be_4$, and the energy of the protons is $E_{kp}= 5Mev$, and the neutrons with the energy E_{kn} of 1.75Mev.

Solution:

The reaction is:

$^7Li_3 + {}^1H_1 \rightarrow {}^7Be_4 + {}^1n_0$

The energy of this reaction is:

Q={(7,016003amu+ 1,007825amu) –(7,016928amu+ 1,008665amu)}.931,5
= 8,023828amu –(8,025593).931,5

Q= -1.644 Mev

By using the equation of the angle, we will obtain:

θ_n = arccos $[E_{kn}(m_n+m_{Be})- E_{kp}(m_{Be}-m_p)-Qm_{Be} / 2\sqrt{(m_n m_p E_{kn} E_{kp})}]$
=arccos $[1,75(1+7)-5(7-1)+7. 1,643 / 2\sqrt{(1. 1. 1,75. 5)}]$
$\theta_n \approx 139,5^0$

235. What the initial mass of $^{235}U_{92}$ is required to operate a 500-MW reactor for one year?

Solution:

The decay constant is:

λ= 0,693 / (7,038.10^8.365,25.24.3600)
λ= 0,693 / (2,221.10^{16})
λ= 3,12.10^{-17} sec^{-1}

The form of this reaction is:

n+ $^{235}U_{92}$ $\longrightarrow$ $^{141}Ba_{56}$+ $^{92}Kr_{36}$+ 3n+ Q
Q= (1,008665amu+ 235,043924amu) – (140,91441amu+
91,926111amu+3(1,008665))amu . 931,5Mev/amu
Q= (236,05258amu– 235,86651amu).931,5Mev/amu
Q= 173,3242Mev.

When:

N= P/ λ.Q

Hence:

N= 500. 10^6 watt / (3,12. 10^{-17}. 173,3242. 10^6)

N= $9,246.10^{16}$ nuclei

When:

$N / m = N_A / A$

$9,246.10^{16} / m = 6,025.10^{23} / 235$

Therefore:

m= $9,246.10^{16}.235 / 6,025.10^{23}$
m = $3,606.10^{-5}$gram.

236. Calculate the energy which released pergram of a fuel for the reaction:
$$^2H_I + ^2H_I \rightarrow ^3H_I + ^1H_I \quad , \quad When \; Q = 4,03Mev$$

Solution:

$N / m = N_A / A$
$N = 6,025.10^{23}.1g / 2$
N= $3,0125.10^{23}$ nuclei

When:

$E = Q. N$

Hence:

E= $4,03.10^6. 3,0125.10^{23}$
E= $1,214.10^{30}$ev

Or:

E= $1,214.10^{30}. 1,6.10^{-19}$
E= $1,942.10^{11}$J.

But the energy which released per agram of Uranium is:

$N / 1g = 6,025.10^{23} / 235$
$N = 2,5638.10^{21}$ nuclei
Q, from the fission is 200Mev

And:

$Q.N = E$
$\quad = 200.10^6. \, 2,5638.10^{21}$
$E \quad = 5,1276.10^{29}$ev

Hence:

$E = 5,1276.10^{29}$ev.$1,6.10^{-19}$
$E = 8,204.10^{10}$J

Therefore:

The energy from the fission is $8,204.10^{10}$J
The energy from the fusion is $1,942.10^{11}$J

237. Show that the energy released when the two deuterium nuclei fuse to form $^3He_2$ with the release of a neutron is 3,27Mev.

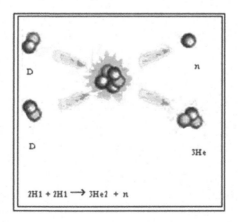

Solution:

The form of the reaction is:

$^2H_1 + \, ^2H_1 \rightarrow \, ^3He_2 + n$

Q = {2(2,014102amu) - (3,016029amu+ 1,008665amu)}. 931,5

Q = {4,028204amu - 4,024694amu}. 931,5Mev/amu

Q = 3,51.10^{-3}.931,5

Q = 3,27Mev.

238. Quant of gamma ray transfer through the mass of 6. 10^{-3} kg of the dry air to get 1,7.10^{12} ions. Each ion has a charge of+e. What the exposure with Roentgen unit?

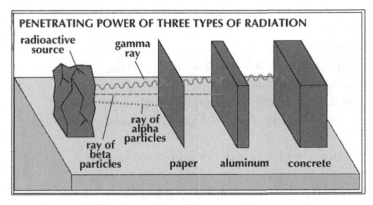

PENETRATING POWER OF THREE TYPES OF RADIATION

http://student.britannica.com/eb/art-53873/The-penetrating-power-of-alpha-rays-beta-rays-and-gamma

Solution:

Exposure(in Roentgens)= The quantity of the charge / mass

= (1/ 2,58.10^{-4}). q / m

= {1/ (1/ 2,58.10^{-4})}. {(1/ 2,58.10^{-4})}/(6.10^{-3})

= 1/ (6.10^{-3})≈166,67 C/kg

Or:

Exposure= 166,67 Roentgen

239. In a biological research institute, a researcher use a rabbit which has mass of 12 kg to mesure the whole body dose from $^{125}I_{53}$ as a radioactive source has a radioactivity of 530 Ci. When 1,2% from gamma rays arrived to this animal and the average of this rays is 0,035Mev with uniforms quantities. 60% of gamma rays reacted in the body of the rabbit. Calculate the whole body dose.

Solution:

When the whole body dose write as:

$$D_W = A_{ac.} \ N. \ \varepsilon. \ f_{frac} \ / \ M$$

When:

A_{ac} is the radioactivity,
N: The number of the particles with some energy
ε: The average of gamma energy,
f_{frac}: The percentage or the fraction of the absorption,
M: The mass of the organ or the tissue which absorb the rays.

The total energy which transfers through the body is:

$$530. \ 3,7.10^{10}. \ 0,035. \ 0,012 = 8,2362.10^9 Mev/s$$

The energy which deposed in the body is:

$$0,6. \ 8,2362.10^9.1,6.10^{-13} J/Mev = 7,9067.10^{-4} \ J/s$$

When:

$$1Gy = 1 J/kg$$

Hence:

The whole body dose D_W is:

$$\{7,9067.10^{-4} \ J/s \ / \ 1J/kg.Gy\}.12kg \ )= 9,488.10^{-3} \ Gy/s$$

Therefore:

$D_W = 9,488.10^{-3}$ Gy/s

240. How much the energy is deposited in the body of a 70 kg adult exposed to a 50 rad dose?

Solution:

When the absorbed dose D= 50/100 rad

$D = 0,5$ J/kg= 0,5 Gy
D= E / m

Hence:

E= D. m= 0,5 J / kg. 70 kg
$E = 35$ Joule.

241. A dose of 500 rem of gamma rays in a short period would be lethal to about half of the people subjected to it. How many rads are in this period?

Solution:

Where D: Absorbed dose, E: effective dose, Q_f weight factor:

And:

E (rem)= D(rad). Q_f

Hence:

500rem= D(rad). 1

When:

$Q_f = 1$ in this case,

Therefore:

D= 500rad.

**242. *A 70 kg person is exposed to 40 mrem of alpha particles (RBE= 12).
Find the absorbed dose from the biological equivalent dose(BED).***

Solution:

Absorbed dose in rad= BED / RBE
$$= 40.10^{-3} \text{ rem} / 12$$
When 1Gy= 100rad

Hence:

Absorbed dose in Grays= $(40.10^{-3} / 12)$rad. $(1/100)$Gy/rad
The energy (E)= (Absorbed dose in grays)(mass)
$$=(40.10^{-5} /12)\text{Gy. (70) kg}$$
E $= 2,23.10^{-3}$ J.

APPENDIXES

APPENDIX A

Periodical Table of Elements

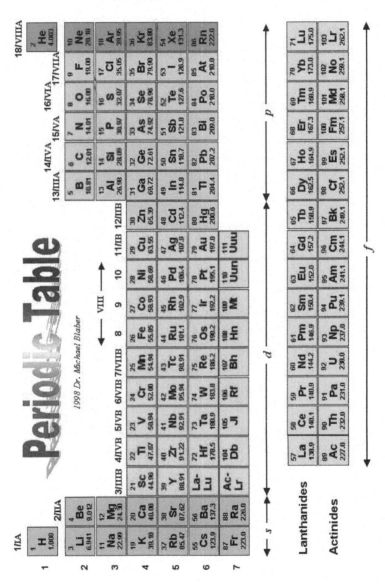

APPENDIX   B

Radiation Dosage

5Sv	50% immediate
1Sv	5% lethal tumours
1/5Sv	0.5% gross genetic birth defect
20mSv	Max.occupational average yearly dose
1mSv	Max.Public average yearly dose
2mSv	Natural backgroung yearly dose

APPENDIX   C

Quantity, Value and Units

Quantity	Value	Unit
Alpha particle mass	$6.644\ 6565.10^{-27}$	Kg
Alpha particle mass in u	$4.001\ 506\ 179\ 149$	Amu
Atomic mass constant	$1.660\ 538\ 86.10^{-27}$	kg
Atomic mass unit-electron volt, relationship	$931.494\ 043.10^{6}$	eV
Atomic mass unit-kilogram, relationship	$1.660\ 538\ 86.10^{-27}$	kg
Atomic unit of energy	$4.359\ 744\ 17.10^{-18}$	J
Avogadro's number N_A	$6.022\ 1367.10^{23}$	mol^{-1}
Bohr magneton	$927.400\ 949.10^{-26}$	JT^{-1}
Bohr radius	$0.529\ 177\ 2108.10^{-10}$	M
Boltzmann's constant k	$1.380\ 658.10^{-23}$	$J\ K^{-1}$
Compton wavelength	$2.426\ 310\ 238.10^{-12}$	m
Constant of conversion from $\hbar c$	197.3	Mev.fm
Deuteron mass	$3.343\ 583\ 35e^{-27}$	Kg
Deuteron mass inamu	$2.013\ 553\ 21270$	Amu
Electron charge magnitude e	$1,60217733.10^{-19}$	C
Electron mass	$9.109\ 3826.10^{-31}$	Kg
Electron mass in u	$5.485\ 799\ 0945.10^{-4}$	amu
Electron mass energy equivalent	$8.187\ 1047.10^{-14}$	J
Electron mass energy equivalentinMev	$0.510\ 998\ 918$	MeV
Electron volt-atomic mass unit Relationship	$1.073\ 544\ 171.10^{-9}$	Amu

Quantity	Value	Unit
Electron volt-joule relationship	$1.602\ 176\ 53.10^{-19}$	J
Electron volt-kilogram relationship	$1.782\ 661\ 81.10^{-36}$	Kg
Fine-structure constant	$7.297\ 352\ 568.10^{-3}$	
Inverse fine-structure constant	$137.035\ 999\ 11$	
Joule-electron volt relationship	$6.241\ 509\ 47.10^{18}$	eV
Joule-atomic mass unit relationship	$6.700\ 5361.10^{9}$	amu
kilogram-atomic mass unit relationship	$6.022\ 1415.10^{26}$	amu
kilogram-electron volt relationship	$5.609\ 588\ 96.10^{35}$	eV
kilogram-joule relationship	$8.987\ 551\ 787....10^{16}$	J
Muon mass	$1.883\ 531\ 40.10^{-28}$	Kg
Muon mass inamu	$0.113\ 428\ 9264$	amu
Muon molar mass	$0.113\ 428\ 9264.10^{-3}$	kg mol^{-1}
Muon Compton wavelength	$11.734\ 441\ 05.10^{-15}$	M
Neutron mass	$1.674\ 927\ 28.10^{-27}$	Kg
Neutron mass inamu	$1.008665\ 915$	amu
Nuclear magneton	$5.050\ 783\ 43.10^{-27}$	J T^{-1}
Planck's constant h	$6,62606891.10^{-34}$	J.s
Proton mass	$1.672\ 621\ 71.10^{-27}$	Kg
Proton mass inamu	$1.00782576\ 466$	amu
Rydberg constant	$10\ 973\ 731.568\ 525$	m^{-1}
Speed of light in vacuum c	$2,99792458.10^{8}$	m/s
Stefan-Boltzmann constant	$5.670\ 400.10^{-8}$	W m^{-2} K^{-4}
Standard atmosphere	$101\ 325$	Pa
Standard acceleration of gravity	$9.806\ 65$	m s^{-2}

Quantity	Value	Unit
Tau mass	3.167 77.10⁻²⁷	Kg
Tau mass inamu	1.907 68	amu
Universal gravitational constant G	6,67259.10⁻¹¹	N.m²/kg²
Universal gas constant R	8,314510	J/(mol. K)

To change	To	Multiply by
atmospheres	cms. of mercury	76
Btu/hour	horsepower	0,0003930
Btu	kilowatt-hour	0,0002931
Btu/hour	watts	0,2931
centimeters	inches	0,3937
centimeters	feet	0,03281
cubic feet	cubic meters	0,0283
cubic meters	cubic feet	35,3145
cubic meters	cubic yards	1,3079
cubic yards	cubic meters	0,7646
degrees	radians	0,01745
dynes	grams	0,00102
feet	meters	0,3048
feet	miles (nautical)	0,0001645
feet	miles (statute)	0,0001894
feet/second	miles/hour	0,6818
hours	days	0,04167
inches	millimeters	25,4000
inches	centimeters	2,5400
kilograms	pounds (avdp or troy)	2,2046
kilometers	miles	0,6214
kilowatt-hour	Btu	3412
knots	nautical miles/hour	1,0

To change	To	Multiply by
knots	statute miles/hour	1,151
meters	feet	3,2808
meters	miles	0,0006214
meters	yards	1,0936
miles	kilometers	1,6093
miles	feet	5280
miles (nautical)	miles (statute)	1,1516
miles (statute)	miles (nautical)	0,8684
miles/hour	feet/minute	88
millimeters	inches	0,0394
pounds (ap or troy)	kilograms	0,3732
pounds (avdp)	kilograms	0,4536
pounds	ounces	16
quarts (dry)	liters	1,1012
quarts (liquid)	liters	0,9463
radians	degrees	57,30
square feet	square meters	0,0929
square kilometers	square miles	0,3861
square meters	square feet	10,7639
square meters	square yards	1,1960
square miles	square kilometers	2,5900
square yards	square meters	0,8361
tons (long)	metric tons	1,016
tons (short)	metric tons	0,9072
tons (long)	pounds	2240
tons (short)	pounds	2000
watts	Btu/hour	3,4121
watts	horsepower	0,001341
yards	meters	0,9144
yards	miles	0,0005682

http://www.infoplease.com/ipa/A0001729.html

SI prefixes for decimal multiples
There are 20 internationally accepted prefixes to denote 10^n multiples of units. 10 of these prefixes denote (n>0) multiples, and other 10 denote (n<0) submultiples:

Number	Factor	Name	Symbol
SI Prefixes for decimal multiples			
1 000 000 000 000 000 000 000 000	10^{24}	yotta	Y
1 000 000 000 000 000 000 000	10^{21}	zetta	Z
1 000 000 000 000 000 000	10^{18}	exa	E
1 000 000 000 000 000	10^{15}	peta	P
1 000 000 000 000	10^{12}	tera	T
1 000 000 000	10^{9}	giga	G
1 000 000	10^{6}	mega	M
1 000	10^{3}	kilo	k
100	10^{2}	hecto	h
10	10^{1}	deca	da
0.1	10^{-1}	deci	d
0.01	10^{-2}	centi	c
0.001	10^{-3}	milli	m
0.000 001	10^{-6}	micro	μ
0.000 000 001	10^{-9}	nano	n
0.000 000 000 001	10^{-12}	pico	p
0.000 000 000 000 001	10^{-15}	femto	f
0.000 000 000 000 000 001	10^{-18}	atto	a
0.000 000 000 000 000 000 001	10^{-21}	zepto	z
0.000 000 000 000 000 000 000 001	10^{-24}	yocto	y

http://members.optus.net/alexey/prefSI.xhtml

Table 1. SI base units

Base quantity	SI base unit	
	Name	Symbol
length	Meter	m
Mass	kilogram	kg
Time	Second	s
electric current	Ampere	A
thermodynamic temperature	Kelvin	K
amount of substance	Mole	mol
luminous intensity	Candela	cd

Dr. Mouaiyad M.S.Alabed

Table 2. Examples of SI derived units

SI derived unit

Derived quantity	Name	Symbol
area	square meter	m^2
volume	cubic meter	m^3
speed, velocity	meter per second	m/s
acceleration	meter per second squared	m/s^2
wave number	reciprocal meter	m^{-1}
mass density	kilogram per cubic meter	kg/m^3
specific volume	cubic meter per kilogram	m^3/kg
current density	ampere per square meter	A/m^2
magnetic field strength	ampere per meter	A/m
amount-of-substance concentration	mole per cubic meter	mol/m^3
luminance	candela per square meter	cd/m^2
mass fraction	kilogram per kilogram, which may be represented by the number 1	kg/kg= 1

Table 3. SI derived units with special names and symbols

SI derived unit

Derived quantity	Name	Symbol	Expression in terms of other SI units	Expression in terms of SI base units
plane angle	radian [a]	rad	-	$m \cdot m^{-1} = 1$ [b]
solid angle	steradian [a]	sr [c]	-	$m^2 \cdot m^{-2} = 1$ [b]
frequency	Hertz	Hz	-	s^{-1}
force	newton	N	-	$m \cdot kg \cdot s^{-2}$
pressure, stress	Pascal	Pa	N/m^2	$m^{-1} \cdot kg \cdot s^{-2}$
energy, work, quantity of heat	Joule	J	$N \cdot m$	$m^2 \cdot kg \cdot s^{-2}$
power, radiant flux	Watt	W	J/s	$m^2 \cdot kg \cdot s^{-3}$
electric charge, quantity of electricity	coulomb	C	-	$s \cdot A$

246

Derived quantity	Name	Symbol	Expression in terms of other SI units	Expression in terms of SI base units
electric potential difference, electromotive force	Volt	V	W/A	$m^2 \cdot kg \cdot s^{-3} \cdot A^{-1}$
capacitance	Farad	F	C/V	$m^{-2} \cdot kg^{-1} \cdot s^4 \cdot A^2$
electric resistance	Ohm	Ω	V/A	$m^2 \cdot kg \cdot s^{-3} \cdot A^{-2}$
electric conductance	siemens	S	A/V	$m^{-2} \cdot kg^{-1} \cdot s^3 \cdot A^2$
magnetic flux	Weber	Wb	V·s	$m^2 \cdot kg \cdot s^{-2} \cdot A^{-1}$
plane angle				
solid angle				
frequency				
force				
Electric resistance				
Electric conductance				
magnetic flux				
magnetic flux density	Tesla	T	Wb/m^2	$kg \cdot s^{-2} \cdot A^{-1}$
inductance	Henry	H	Wb/A	$m^2 \cdot kg \cdot s^{-2} \cdot A^{-2}$
Celsius temperature	degree Celsius	°C	-	K
luminous flux	Lumen	lm	cd·sr [c]	$m^2 \cdot m^{-2} \cdot cd = cd$
illuminance	Lux			$m^2 \cdot m^{-4} \cdot cd = m^2 \cdot cd$
activity (of a radionuclide)	Becquerel	Bq	-	s^{-1}
absorbed dose, specific energy (imparted), kerma	Gray	Gy	J/kg	$m^2 \cdot s^{-2}$
dose equivalent [d]	Sievert	Sv	J/kg	$m^2 \cdot s^{-2}$
catalytic activity	Katal	kat		$s^{-1} \cdot mol$
surface tension	newton per meter			N/m
angular velocity	radian per second			rad/s
angular acceleration	radian per second squared			rad/s^2
heat flux density, irradiance	watt per square meter			W/m^2

247

Derived quantity	Name	Symbol	Expression in terms of other SI units	Expression in terms of SI base units
heat capacity, entropy	joule per kelvin			J/K
specific heat capacity, specific entropy	joule per kilogram kelvin			J/(kg·K)
specific energy	joule per kilogram			J/kg
thermal conductivity	watt per meter kelvin			W/(m·K)
energy density	joule per cubic meter			J/m³
electric field strength	volt per meter			V/m
electric charge density	coulomb per cubic meter			C/m³
electric flux density	coulomb per square meter			C/m²
permittivity	farad per meter			F/m
permeability	henry per meter			H/m
molar energy	joule per mole			J/mol
molar entropy, molar heat capacity	joule per mole kelvin			J/(mol·K)
exposure (x and γ rays)	coulomb per kilogram			C/kg
absorbed dose rate	gray per second			Gy/s
radiant intensity	watt per steradian			W/sr
radiance	watt per square meter steradian			W/(m²·sr)
catalytic (activity) concentration	katal per cubic meter			kat/m³

Table 4. Examples of SI derived units whose names and symbols include SI derived units with special names and symbols

	SI derived unit	
Derived quantity	Name	Symbol
dynamic viscosity	pascal second	Pa·s
moment of force	newton meter	N·m
surface tension	newton per meter	N/m
angular velocity	radian per second	rad/s
angular acceleration	radian per second squared	rad/s^2
heat flux density, irradiance	watt per square meter	W/m^2
heat capacity, entropy	joule per kelvin	J/K
specific heat capacity, specific entropy	joule per kilogram kelvin	J/(kg·K)
specific energy	joule per kilogram	J/kg
thermal conductivity	watt per meter kelvin	W/(m·K)
energy density	joule per cubic meter	J/m^3
electric field strength	volt per meter	V/m
electric charge density	coulomb per cubic meter	C/m^3
electric flux density	coulomb per square meter	C/m^2
Permittivity	farad per meter	F/m
Permeability	henry per meter	H/m
molar energy	joule per mole	J/mol
molar entropy, molar heat capacity	joule per mole kelvin	J/(mol·K)
exposure (x and γrays)	coulomb per kilogram	C/kg
absorbed dose rate	gray per second	Gy/s
radiant intensity	watt per steradian	W/sr
Radiance	watt per square meter steradian	W/(m^2·sr)
catalytic (activity) concentration	katal per cubic meter	kat/m^3

http://physics.nist.gov/cuu/Units/units.html

Capital	Low-case	Greek Name	English
A	α	Alpha	a
B	β	Beta	b
Γ	γ	Gamma	g
Δ	δ	Delta	d
E	ε	Epsilon	e
Z	ζ	Zeta	z
H	η	Eta	h
Θ	θ	Theta	th
I	ι	Iota	i
K	κ	Kappa	k
Λ	λ	Lambda	l
M	μ	Mu	m
N	ν	Nu	n
Ξ	ξ	Xi	x
O	o	Omicron	o
Π	π	Pi	p
P	ρ	Rho	r
Σ	σ	Sigma	s
T	τ	Tau	t
Υ	υ	Upsilon	u
Φ	ϕ	Phi	ph
X	χ	Chi	ch
Ψ	ψ	Psi	ps
Ω	ω	Omega	o

http://www.physlink.com/Reference/GreekAlphabet.cfm

List of Elements in Atomic Number Order.
Name symbol atomic wt atomic number state half-life

At No	Symbol	Name		Atomic Wt	Half-life
					10,4min
0	n	Neutron	1		1,008665
	1H	Hydrogen	1	stable	1.0078250
1	2H	Deuterium	2	stable	2,014102
	3H	Tritium	3	12.33yr	3,016049
		Helium	3	3,016029	
			4	4.002602	
2	He		6	806,7msec	6.0188880
			7	7.0280302	
			8	119msec	8.0339218
3	Li	Lithium	6	6,015121	
			7	7,016003	
4	Be	Beryllium	7	53,29days	7,016928
			9	9,012182	
5	B	Boron	10	10.012936	
			11	11,009305	
		Carbon	11	20,385min	11,011433
6	C		12	12,000000	
			13	13,003355	
			14	5730yr	14,003242
		Nitrogen	13	9,965min	13,005738
7	N		14	15,000108	
			15	14.003074	
		Oxygen	15	122,24s	15,003065
8	O		16	15,994915	
			18	17,999160	
9	F	Fluorine	19	18,998404	
10	Ne	Neon	20	19,992435	
			22	21,991383	
		Sodium	22	2,6088yr	21,994434
11	Na		23	22,989767	
			24	14,9590h	23,990961
12	Mg	Magnesium	24	23,985042	
13	Al	Aluminium	27		26,981538
14	Si	Silicon	28	30,975362	27,976927
			31	157,3min	

15	P	Phosphorus	31	30,973762	
			32	14,262days	31,973908
16	S	Sulfur	32	31,972071	
			35	87,51days	34,969033
17	Cl	Chlorine	35	34,968853	
			37	36,965903	
18	Ar	Argon	40	39.962384	
19	K	Potassium	39	38,963708	
			40	1,227.10⁹yr	39,964000
20	Ca	Calcium	40	39,962591	
21	Sc	Scandium	45	44,955911	
22	Ti	Titanium	48	47,947947	
23	V	Vanadium	51	50.943962	
24	Cr	Chromium	52	51,940511	
25	Mn	Manganese	55	54,938048	
26	Fe	Iron	56	55,934940	
27	Co	Cobalt	59	58,933198	
			60	5,2714yr	59,933820
28	Ni	Nickel	58	59,930789	57,935346
			60		
29	Cu	Copper	63	64,927791	
			65	62,929599	
30	Zn	Zinc	64	63,929144	
			66	65,926035	
31	Ga	Gallium	69	68,925580	
32	Ge	Germanium	72	71,922079	
			74	73,921177	
33	As	Arsenic	75	74,921594	
34	Se	Selenium	80	79,916519	
35	Br	Bromine	79	78,918336	
36	Kr	Krypton	84	83,911508	
37	Rb	Rubidium	85	84,911793	
38	Sr	Strontium	86	85,909266	
			88	87,905618	
			90	29,1yr	89,907737
39	Y	Yttrium	89	88,905847	
40	Zr	Zirconium	90	89,904702	
41	Nb	Niobium	93	92.906376	
42	Mo	Molybdenum	98	97,905407	
43	Tc	Technetium	98	4,2.106yr	97,907215
44	Ru	Ruthenium	102	101,904348	
45	Rh	Rhodium	103	102.905502	
46	Pd	Palladium	106	105,903481	

47	Ag	Silver	107	106,905091	
			109	108,904754	
48	Cd	Cadmium	114	113,903359	
49	In	Indium	115	$4,41.10^{14}$yr	114,903876
50	Sn	Tin	120	119,902197	
51	Sb	Antimony	121	120,903820	
52	Te	Tellurium	130	$1,25.10^{21}$yr>	129,906228
53	I	Iodine	127	126.904474	
			131	8,04days	130,906111
54	Xe	Xenon	132	131,904141	
			136	$2,36.10^{21}$y$\leq$	135,90721r
55	Cs	Caesium	133	132.905436	
56	Ba	Barium	137	136,905816	
			138	137,905236	
57	La	Lanthanum	139	138.906346	
58	Ce	Cerium	140	139,905434	
59	Pr	Praseodymium	141	140.907647	
60	Nd	Neodymium	142	141,907718	
61	Pm	Promethium	145	17,7yr	144,912745
62	Sm	Samarium	152	151,919728	
63	Eu	Europium	153	152,921226	
64	Gd	Gadolinium	158	157,924099	
65	Tb	Terbium	159	158.925344	
66	Dy	Dysprosium	164	163,929172	
67	Ho	Holmium	165	164.930320	
68	Er	Erbium	166	165,930292	
69	Tm	Thulium	169	168.934213	
70	Yb	Ytterbium	174	173,938861	
71	Lu	Lutetium	175	174.940772	
72	Hf	Hafnium	180	179,946547	
73	Ta	Tantalum	181	180.947993	
74	W	Tungsten	184	3.10^{17}yr<	183,950929
75	Re	Rhenium	187	$4,35.10^{10}$yr	186,955746
76	Os	Osmium	191	15,4days	190,960922
			192	191,961468	
77	Ir	Iridium	191	190,960585	
			193	192,962916	
78	Pt	Platinum	195	194,964765	
79	Au	Gold	197		
			198	2.69days	196.966543
80	Hg	Mercury	199	198,968253	
			202	201,970617	
81	Tl	Thallium	205	204,974400	

82	Pb	Lead	206	205,974440	
			207	206,975871	
			208	207,976627	
			210	22,3yr	209,984163
			211	36,1min	210,988734
			212	10,64h	211,991872
			214	26,8min	213,999798
83	Bi	Bismuth	209	208.980374	
			211	2,14min	210,987254
84	Po	Polonium	210	138,376days	209,982848
			214	164,3μs	213,995177
85	At	Astatine	218	1,6s	218,00868
86	Rn	Radon	222	3,8235days	222,017571
87	Fr	Francium	223	21,8min	223,019733
88	Ra	Radium	226	1600yr	226,025402
89	Ac	Actinium	227	21,773yr	227,027749
90	Th	Thorium	228	1,9131yr	228,028716
			232	$1,405.10^{10}$yr	232,038051
91	Pa	Protactinium	231	$3,276.10^4$yr	231.035880
92	U	Uranium	232	68,9yr	232,037131
			233	$1,592.10^5$yr	233,039630
			235	$7,038.10^8$yr	235,043924
			236	$2,3415.10^7$yr	236,045562
			238	$4,468.10^9$yr	238,050784
			239	23,50min	239,054289
93	Np	Neptunium	239	2,355days	239,052932
94	Pu	Plutonium	239	24,119yr	239,052157
95	Am	Americium	243	7380yr	243,061373
96	Cm	Curium	245	8500yr	245,065484
97	Bk	Berkelium	247	1380yr	147,07030
98	Cf	Californium	249	351yr	249,074844
99	Es	Einsteinium	254	275,7days	254,08802
100	Fm	Fermium	253	3 days	253,085174
101	Md	Mendelevium	255	27min	255,09107
102	No	Nobelium	255	3,1min	255,09324
103	Lr	Lawrencium	257	0,646s	257,0995
104	Rf	Rutherfordium	261	65s	261,1086
105	Db	Dubnium	262	34s	262,1138
106	Sg	Seaborgium	263	0,8s	263,1182
107	Bh	Bohrium	262	102ms	262,1231
108	Hs	Hassium	264	0,08ms	264,1285
109	Mt	Meitnerium	266	3,4ms	266,1378

RELATIONS AND FORMULAS IN THIS BOOK!

$$E = (P^2c^2 + m^2c^4)^{1/2} \sim (10^{-21} + 10^{-26})^{1/2}J$$
$$\rho_{ch}(r) = \rho^0_{ch} / \{1 + e^{(r-R)/a}\}$$

$$= \lambda = h / p$$
$$= hc / pc =$$
$$= 2\pi \hbar c / pc$$
$$\rho(R) = \rho(0)/2$$
$$R = r_0 A^{1/3}$$
$$\rho = M(A,Z) / V(A,Z) \approx m_N A /(4/3) \pi R^3$$
$$= 3m_N A / \{4\pi r_0^3 A\} =$$
$$= 3m_N / 4\pi r_0^3$$
$$\rho \approx 2 \cdot 10^{14} \text{ g/cm}^3$$
$$E_0(A,Z) = Zm_p + (A-Z)m_n - M_N(A,Z)$$
$$M(A,Z) = M_N(A,Z) + Zm_e$$
$$E_{BE} = ZM(^1H) + (A-Z)m_n - M(A,Z)$$
$$E_{BE} = ZM(^1H) + (A-Z)m_n - M(A,Z) = Z\Delta(^1H) + (A-Z)\Delta_n - \Delta(A,Z)$$
$$B_n = M(A-1,Z) + m_n - M(A,Z) = \Delta(A-1,Z) + \Delta_n - \Delta(A,Z)$$
$$B_p = M(A-1,Z-1) + M(^1H) - M(A,Z) = \Delta(A-1,Z-1) + \Delta(^1H) - \Delta(A,Z)$$
$$E_s = a_2 A^{2/3}(1 + 2\varepsilon^2 /5 + ...)$$
$$E_c = a_3 Z^2 A^{-1/3}(1 - \varepsilon^2 /5 + ...)$$
$$\Delta E = -\varepsilon^2/5(2a_2 A^{2/3} - a_3 \cdot Z^2 A^{-1/3})$$
$$Z^2/A \geq 2a_2/a_3 \approx 48$$
$$X \rightarrow A + B + (C + ...)$$
$$X + Y \rightarrow A + B + (C + ...)$$

$$\Sigma E = \text{Const}$$
$$\Sigma P = \text{Const}$$
$$\Sigma j` = \text{Const}$$
$$\Sigma Q = \text{Const}$$
$$dN(t) = -\lambda N(t)dt$$
$$N(t) = N(0) \exp(-\lambda t) = N(0) \exp(-t/\tau)$$

$N(T_{1/2})= N(0) /2= N(0) \exp(-\lambda T_{1/2})$

$\ln2= \lambda T_{1/2} ; T_{1/2}= \ln2/\lambda= \tau\ln2$

$P^{\backslash}_A+ P^{\backslash}_B = 0$

$E_{kB}= \Delta M \cdot M_A / (M_A+ M_B);$

$N(t)= N(0) \exp(-t/\tau)$

$N(t)= N(0)/2= N(0) \exp(-t/\tau)$

$t = t_0 / \sqrt{(1-(v^2/c^2))}$

$t = \tau \ln2 / \sqrt{(1-(v^2/c^2))}$

$P = [\sqrt{(E_k^{2+} 2E_k mc^2)}] / c$

$P = mv / \sqrt{(1-(v^2/c^2))}$

$P = [\sqrt{(E_k^{2+} 2 E_k mc^2)}]. [\sqrt{(1-(v^2/c^2))}] / mc$

$l = vt$

$^{14}C \rightarrow ^{14}N+ e^-+ \tilde{\upsilon}_e$ (β⁻)

$^{11}C \rightarrow ^{11}B+ e^++ \upsilon_e$ (β⁺)

(4.13)

$^7Be+ e^- \rightarrow ^7Li+ \upsilon_e$ (e)

$M_N(Z, A)= M_N(Z+ 1, A)+ m_e+ T_N+ T_e+ E_v$

$M(Z, A)= M_N(Z+ 1, A)+ Z \cdot m_e$

$M(Z, A)= M(Z+ 1, A)+ T_N+ T_e+ E_v$

$M(Z, A)= M(Z - 1, A)+2 m_e+ T_N+ T_e+ E_v$

$M(Z, A)= M(Z - 1, A)+ T_N+ E_v$

$W= \int_0^{RN} |\Psi(r)|^2 r^2 dr \ll 1$

$^{60}Co \rightarrow ^{60}Ni+ e^-+ \tilde{\upsilon}_e$

$J^{\backslash}_i = J^{\backslash}_f+s^{\backslash}_{e+}s^{\backslash}_\upsilon+l^{\backslash}_e+\upsilon$

$l= Rpc/\hbar c= RT/\hbar c \leq 1/10 \ll 1$

$A+ B \rightarrow a+ b+ c+........$

$E_{kA}+ M_A+M_B= \Sigma m_f+ \Sigma E_{kf}, P^{\backslash}_A= \Sigma P^{\backslash}_f$

$E^{'}_{kA}+ E^{'}_{kB}+M_A+ M_B= \Sigma m_f+ \Sigma E^{'}_{kf}, P^{\wedge}_A+ P^{\wedge}_B= \Sigma P^{\wedge}_f= 0$

$(E_{kA}+ M_A+M_B)^2 - P^2_A= (\Sigma m_f)^2$

$E_{KA}= (1/2M_B) [(\Sigma m_f)^2-(\Sigma m_i)^2]=(1/2M_B)[(\Sigma m_f-\Sigma m_i)(\Sigma m_f+\Sigma m_i)]$

$E_{KA}= |Q| [1+(M_A / M_B)+(|Q| / 2M_B)]$

$d\sigma/d\theta= (dN/d\theta)/I.n$

$[d\sigma/d\theta]= [cm^2/sterad]$

$\sigma = \int(d\sigma/d\theta) d\theta$

$\sigma_{tot}= \sigma_{abs}+ \sigma_{el}= 2\pi R^2$

$dN(t) = In\sigma dt - \lambda N(t)dt$

$N(t) = In\sigma (1-e^{-\lambda t})$

β^- disitegration

β^+-disintegration -

e- capture -

$Q_\beta^- = \Delta (A,Z) - \Delta(A,Z+1)$

$Q_{\beta+} = \Delta (A,Z) - \Delta(A,Z-1) - 2m_ec^2$

$Q_e = \Delta (A,Z) - \Delta(A,Z-1)$

$E = (p^2c^2 + m^2c^4)^{1/2} = E_k + mc^2$

$p = [(E_k^2 + 2 E_k \cdot mc^2)^{1/2}]/c$

$E^2 = p^2 + m^2 = (E_k + m)^2$

$p^2 = E_k^2 + 2 E_k m$

$E_{bind} = a_1 A - a_2 A^{2/3} - a_3 Z^2/A^{1/3} - a_4(A-2Z)^2/A + \delta A^{-3}$

$h = 6{,}62606 . 10^{-34}$ Js

$1ev = 1{,}602176 . 10^{-19}$ J

$h = 6{,}62606 . 10^{-34}$ Js (ev / $1{,}602176 . 10^{-19}$ J)

$\quad = 4{,}13566 . 10^{-21Mev}$.sec

therefore:

$\hbar = 6{,}582 . 10^{-22}$ Mev.sec

Energy - $1Mev = 10^6$ eV $= 10^{-3}$

$Gev = 1{,}6 . 10^{-13}$ J.

Mass - $1Mev/c^2$,

and also $1u = M(^{12}C)/12 = 931.5 Mev/c^2 =$

$\quad\quad\quad = 1.66 \; 10^{-24}$ g

The length - 1 fm $= 10^{-13}$cm $= 10^{-15}$ m

Dr. Mouaiyad M.S.Alabed

Isotopes for elements 1-15

Listing of known isotopes

n ↓ \ p →	1 H	2 He	3 Li	4 Be	5 B	6 C	7 N	8 O	9 F	10 Ne	11 Na	12 Mg	13 Al	14 Si	15 P
0	^{1}H	^{2}He	Li	Be	5	6									
1	^{2}D	^{3}He	^{4}Li	^{5}Be	B	C	7								
2	^{3}T	^{4}He	^{5}Li	^{6}Be	^{7}B	^{8}C	N	8							
3	^{4}H	^{5}He	^{6}Li	^{7}Be	^{8}B	^{9}C	^{10}N	O	9	10					
4	^{5}H	^{6}He	^{7}Li	^{8}Be	^{9}B	^{10}C	^{11}N	^{12}O	F	Ne					
5	^{6}H	^{7}He	^{8}Li	^{9}Be	^{10}B	^{11}C	^{12}N	^{13}O	^{14}F	^{15}Ne	11				
6	^{7}H	^{8}He	^{9}Li	^{10}Be	^{11}B	^{12}C	^{13}N	^{14}O	^{15}F	^{16}Ne	Na	12	13	14	
7		^{9}He	^{10}Li	^{11}Be	^{12}B	^{13}C	^{14}N	^{15}O	^{16}F	^{17}Ne	^{18}Na	Mg	Al	Si	15
8		^{10}He	^{11}Li	^{12}Be	^{13}B	^{14}C	^{15}N	^{16}O	^{17}F	^{18}Ne	^{19}Na	^{20}Mg	^{21}Al	^{22}Si	P
9			^{12}Li	^{13}Be	^{14}B	^{15}C	^{16}N	^{17}O	^{18}F	^{19}Ne	^{20}Na	^{21}Mg	^{22}Al	^{23}Si	^{24}P
10				^{14}Be	^{15}B	^{16}C	^{17}N	^{18}O	^{19}F	^{20}Ne	^{21}Na	^{22}Mg	^{23}Al	^{24}Si	^{25}P
11					^{16}B	^{17}C	^{18}N	^{19}O	^{20}F	^{21}Ne	^{22}Na	^{23}Mg	^{24}Al	^{25}Si	^{26}P
12					^{17}B	^{18}C	^{19}N	^{20}O	^{21}F	^{22}Ne	^{23}Na	^{24}Mg	^{25}Al	^{26}Si	^{27}P
13					^{18}B	^{19}C	^{20}N	^{21}O	^{22}F	^{23}Ne	^{24}Na	^{25}Mg	^{26}Al	^{27}Si	^{28}P
14					^{19}B	^{20}C	^{21}N	^{22}O	^{23}F	^{24}Ne	^{25}Na	^{26}Mg	^{27}Al	^{28}Si	^{29}P
15						^{21}C	^{22}N	^{23}O	^{24}F	^{25}Ne	^{26}Na	^{27}Mg	^{28}Al	^{29}Si	^{30}P
16						^{22}C	^{23}N	^{24}O	^{25}F	^{26}Ne	^{27}Na	^{28}Mg	^{29}Al	^{30}Si	^{31}P
17							^{24}N	^{25}O	^{26}F	^{27}Ne	^{28}Na	^{29}Mg	^{30}Al	^{31}Si	^{32}P
18								^{26}O	^{27}F	^{28}Ne	^{29}Na	^{30}Mg	^{31}Al	^{32}Si	^{33}P
19									^{28}F	^{29}Ne	^{30}Na	^{31}Mg	^{32}Al	^{33}Si	^{34}P
20									^{29}F	^{30}Ne	^{31}Na	^{32}Mg	^{33}Al	^{34}Si	^{35}P
21										^{31}Ne	^{32}Na	^{33}Mg	^{34}Al	^{35}Si	^{36}P
22										^{32}Ne	^{33}Na	^{34}Mg	^{35}Al	^{36}Si	^{37}P
23											^{34}Na	^{35}Mg	^{36}Al	^{37}Si	^{38}P
24											^{35}Na	^{36}Mg	^{37}Al	^{38}Si	^{39}P
25												^{37}Mg	^{38}Al	^{39}Si	^{40}P
26												^{38}Mg	^{39}Al	^{40}Si	^{41}P
27													^{40}Al	^{41}Si	^{42}P
28													^{41}Al	^{42}Si	^{43}P
29															^{44}P
30															^{45}P
31															^{46}P

Isotopes for elements 16-30

n↓ \ p→	16 S	17 Cl	18 Ar	19 K	20 Ca	21 Sc	22 Ti	23 V	24 Cr	25 Mn	26 Fe	27 Co	28 Ni	29 Cu	30 Zn
13	^{29}S														
14	^{30}S	^{31}Cl													
15	^{31}S	^{32}Cl	^{33}Ar												
16	^{32}S	^{33}Cl	^{34}Ar	^{35}K											
17	^{33}S	^{34}Cl	^{35}Ar	^{36}K	^{37}Ca										
18	^{34}S	^{35}Cl	^{36}Ar	^{37}K	^{38}Ca										
19	^{35}S	^{36}Cl	^{37}Ar	^{38}K	^{39}Ca	^{40}Sc	^{41}Ti								
20	^{36}S	^{37}Cl	^{38}Ar	^{39}K	^{40}Ca	^{41}Sc	^{42}Ti								
21	^{37}S	^{38}Cl	^{39}Ar	^{40}K	^{41}Ca	^{42}Sc	^{43}Ti	^{44}V	^{45}Cr						
22	^{38}S	^{39}Cl	^{40}Ar	^{41}K	^{42}Ca	^{43}Sc	^{44}Ti	^{45}V	^{46}Cr						
23	^{39}S	^{40}Cl	^{41}Ar	^{42}K	^{43}Ca	^{44}Sc	^{45}Ti	^{46}V	^{47}Cr	^{48}Mn	^{49}Fe				
24	^{40}S	^{41}Cl	^{42}Ar	^{43}K	^{44}Ca	^{45}Sc	^{46}Ti	^{47}V	^{48}Cr	^{49}Mn	^{50}Fe	^{51}Co			
25	^{41}S	^{42}Cl	^{43}Ar	^{44}K	^{45}Ca	^{46}Sc	^{47}Ti	^{48}V	^{49}Cr	^{50}Mn	^{51}Fe	^{52}Co	^{53}Ni		
26	^{42}S	^{43}Cl	^{44}Ar	^{45}K	^{46}Ca	^{47}Sc	^{48}Ti	^{49}V	^{50}Cr	^{51}Mn	^{52}Fe	^{53}Co	^{54}Ni	^{55}Cu	
27	^{43}S	^{44}Cl	^{45}Ar	^{46}K	^{47}Ca	^{48}Sc	^{49}Ti	^{50}V	^{51}Cr	^{52}Mn	^{53}Fe	^{54}Co	^{55}Ni	^{56}Cu	^{57}Zn
28	^{44}S	^{45}Cl	^{46}Ar	^{47}K	^{48}Ca	^{49}Sc	^{50}Ti	^{51}V	^{52}Cr	^{53}Mn	^{54}Fe	^{55}Co	^{56}Ni	^{57}Cu	^{58}Zn
29	^{45}S	^{46}Cl	^{47}Ar	^{48}K	^{49}Ca	^{50}Sc	^{51}Ti	^{52}V	^{53}Cr	^{54}Mn	^{55}Fe	^{56}Co	^{57}Ni	^{58}Cu	^{59}Zn
30				^{49}K	^{50}Ca	^{51}Sc	^{52}Ti	^{53}V	^{54}Cr	^{55}Mn	^{56}Fe	^{57}Co	^{58}Ni	^{59}Cu	^{60}Zn
31				^{50}K			^{53}Ti	^{54}V	^{55}Cr	^{56}Mn	^{57}Fe	^{58}Co	^{59}Ni	^{60}Cu	^{61}Zn
32								^{55}V	^{56}Cr	^{57}Mn	^{58}Fe	^{59}Co	^{60}Ni	^{61}Cu	^{62}Zn
33									^{57}Cr	^{58}Mn	^{59}Fe	^{60}Co	^{61}Ni	^{62}Cu	^{63}Zn
34										^{59}Mn	^{60}Fe	^{61}Co	^{62}Ni	^{63}Cu	^{64}Zn
35											^{61}Fe	^{62}Co	^{63}Ni	^{64}Cu	^{65}Zn
36											^{62}Fe	^{63}Co	^{64}Ni	^{65}Cu	^{66}Zn
37												^{64}Co	^{65}Ni	^{66}Cu	^{67}Zn
38													^{66}Ni	^{67}Cu	^{68}Zn
39													^{67}Ni	^{68}Cu	^{69}Zn
40													^{68}Ni	^{69}Cu	^{70}Zn
41														^{70}Cu	^{71}Zn
42															^{72}Zn
43															^{73}Zn
44															^{74}Zn
45															^{75}Zn
46															^{76}Zn
47															^{77}Zn

Isotopes for elements 31-45

p → 31
n ↓ Ga 32

n	Ga (31)	Ge (32)	As (33)	Se (34)	Br (35)	Kr (36)	Rb (37)	Sr (38)	Y (39)	Zr (40)	Nb (41)	Mo (42)	Tc (43)	Ru (44)	Rh (45)
28	^{59}Ga	Ge													
29	^{60}Ga	^{61}Ge													
30	^{61}Ga														
31	^{62}Ga		33												
32	^{63}Ga	^{64}Ge	As	34											
33	^{64}Ga	^{65}Ge		Se	35	36									
34	^{65}Ga	^{66}Ge		^{68}Se	Br	Kr	37								
35	^{66}Ga	^{67}Ge	^{68}As	^{69}Se		Rb									
36	^{67}Ga	^{68}Ge	^{69}As	^{70}Se		^{72}Kr	38								
37	^{68}Ga	^{69}Ge	^{70}As	^{71}Se	^{72}Br	^{73}Kr	^{74}Rb	Sr	39						
38	^{69}Ga	^{70}Ge	^{71}As	^{72}Se	^{73}Br	^{74}Kr	^{75}Rb		Y	40					
39	^{70}Ga	^{71}Ge	^{72}As	^{73}Se	^{74}Br	^{75}Kr	^{76}Rb	^{77}Sr		Zr					
40	^{71}Ga	^{72}Ge	^{73}As	^{74}Se	^{75}Br	^{76}Kr	^{77}Rb	^{78}Sr							
41	^{72}Ga	^{73}Ge	^{74}As	^{75}Se	^{76}Br	^{77}Kr	^{78}Rb	^{79}Sr		^{81}Zr	41				
42	^{73}Ga	^{74}Ge	^{75}As	^{76}Se	^{77}Br	^{78}Kr	^{79}Rb	^{80}Sr	^{81}Y	^{82}Zr	Nb	42			
43	^{74}Ga	^{75}Ge	^{76}As	^{77}Se	^{78}Br	^{79}Kr	^{80}Rb	^{81}Sr	^{82}Y	^{83}Zr	^{84}Nb	Mo			
44	^{75}Ga	^{76}Ge	^{77}As	^{78}Se	^{79}Br	^{80}Kr	^{81}Rb	^{82}Sr	^{83}Y	^{84}Zr			43		
45	^{76}Ga	^{77}Ge	^{78}As	^{79}Se	^{80}Br	^{81}Kr	^{82}Rb	^{83}Sr	^{84}Y	^{85}Zr	^{86}Nb	^{87}Mo	Tc		
46	^{77}Ga	^{78}Ge	^{79}As	^{80}Se	^{81}Br	^{82}Kr	^{83}Rb	^{84}Sr	^{85}Y	^{86}Zr	^{87}Nb	^{88}Mo		44	
47	^{78}Ga	^{79}Ge	^{80}As	^{81}Se	^{82}Br	^{83}Kr	^{84}Rb	^{85}Sr	^{86}Y	^{87}Zr	^{88}Nb	^{89}Mo	^{90}Tc	Ru	45
48	^{79}Ga	^{80}Ge	^{81}As	^{82}Se	^{83}Br	^{84}Kr	^{85}Rb	^{86}Sr	^{87}Y	^{88}Zr	^{89}Nb	^{90}Mo	^{91}Tc	^{92}Ru	Rh
49	^{80}Ga	^{81}Ge	^{82}As	^{83}Se	^{84}Br	^{85}Kr	^{86}Rb	^{87}Sr	^{88}Y	^{89}Zr	^{90}Nb	^{91}Mo	^{92}Tc	^{93}Ru	
50	^{81}Ga	^{82}Ge	^{83}As	^{84}Se	^{85}Br	^{86}Kr	^{87}Rb	^{88}Sr	^{89}Y	^{90}Zr	^{91}Nb	^{92}Mo	^{93}Tc	^{94}Ru	^{95}Rh
51	^{82}Ga	^{83}Ge	^{84}As	^{85}Se	^{86}Br	^{87}Kr	^{88}Rb	^{89}Sr	^{90}Y	^{91}Zr	^{92}Nb	^{93}Mo	^{94}Tc	^{95}Ru	^{96}Rh
52	^{83}Ga	^{84}Ge	^{85}As	^{86}Se	^{87}Br	^{88}Kr	^{89}Rb	^{90}Sr	^{91}Y	^{92}Zr	^{93}Nb	^{94}Mo	^{95}Tc	^{96}Ru	^{97}Rh
53			^{86}As	^{87}Se	^{88}Br	^{89}Kr	^{90}Rb	^{91}Sr	^{92}Y	^{93}Zr	^{94}Nb	^{95}Mo	^{96}Tc	^{97}Ru	^{98}Rh
54			^{87}As	^{88}Se	^{89}Br	^{90}Kr	^{91}Rb	^{92}Sr	^{93}Y	^{94}Zr	^{95}Nb	^{96}Mo	^{97}Tc	^{98}Ru	^{99}Rh
55				^{89}Se	^{90}Br	^{91}Kr	^{92}Rb	^{93}Sr	^{94}Y	^{95}Zr	^{96}Nb	^{97}Mo	^{98}Tc	^{99}Ru	^{100}Rh
56					^{91}Br	^{92}Kr	^{93}Rb	^{94}Sr	^{95}Y	^{96}Zr	^{97}Nb	^{98}Mo	^{99}Tc	^{100}Ru	^{101}Rh
57				^{91}Se	^{92}Br	^{93}Kr	^{94}Rb	^{95}Sr	^{96}Y	^{97}Zr	^{98}Nb	^{99}Mo	^{100}Tc	^{101}Ru	^{102}Rh
58						^{94}Kr	^{95}Rb	^{96}Sr	^{97}Y	^{98}Zr	^{99}Nb	^{100}Mo	^{101}Tc	^{102}Ru	^{103}Rh
59						^{95}Kr	^{96}Rb	^{97}Sr	^{98}Y	^{99}Zr	^{100}Nb	^{101}Mo	^{102}Tc	^{103}Ru	^{104}Rh
60							^{97}Rb	^{98}Sr	^{99}Y	^{100}Zr	^{101}Nb	^{102}Mo	^{103}Tc	^{104}Ru	^{105}Rh
61						^{97}Kr	^{98}Rb	^{99}Sr	^{100}Y	^{101}Zr	^{102}Nb	^{103}Mo	^{104}Tc	^{105}Ru	^{106}Rh
62							^{99}Rb			^{102}Zr	^{103}Nb	^{104}Mo	^{105}Tc	^{106}Ru	^{107}Rh
63									^{102}Y		^{104}Nb	^{105}Mo	^{106}Tc	^{107}Ru	^{108}Rh
64											^{105}Nb	^{106}Mo	^{107}Tc	^{108}Ru	^{109}Rh
65											^{106}Nb	^{107}Mo	^{108}Tc	^{109}Ru	^{110}Rh
66												^{108}Mo	^{109}Tc	^{110}Ru	^{111}Rh
67													^{110}Tc	^{111}Ru	^{112}Rh
68														^{112}Ru	^{113}Rh
69														^{113}Ru	^{114}Rh

Isotopes for elements 46-60

p → (proton number across), n ↓ (neutron number down)

n↓	46 Pd	47 Ag	48 Cd	49 In	50 Sn	51 Sb	52 Te	53 I	54 Xe	55 Cs	56 Ba	57 La	58 Ce	59 Pr	60 Nd
51	^{97}Pd														
52	^{98}Pd	^{99}Ag	^{100}Cd	49											
53	^{99}Pd	^{100}Ag	^{101}Cd	In	50	51	52								
54	^{100}Pd	^{101}Ag	^{102}Cd		Sn	Sb	Te								
55	^{101}Pd	^{102}Ag	^{103}Cd	^{104}In			^{107}Te								
56	^{102}Pd	^{103}Ag	^{104}Cd	^{105}In			^{108}Te								
57	^{103}Pd	^{104}Ag	^{105}Cd	^{106}In	^{107}Sn		^{109}Te	53	54						
58	^{104}Pd	^{105}Ag	^{106}Cd	^{107}In	^{108}Sn			I	Xe	55					
59	^{105}Pd	^{106}Ag	^{107}Cd	^{108}In	^{109}Sn	^{110}Sb	^{111}Te		^{113}Xe	Cs					
60	^{106}Pd	^{107}Ag	^{108}Cd	^{109}In	^{110}Sn	^{111}Sb	^{112}Te				56				
61	^{107}Pd	^{108}Ag	^{109}Cd	^{110}In	^{111}Sn	^{112}Sb	^{113}Te		^{115}Xe	^{116}Cs	Ba				
62	^{108}Pd	^{109}Ag	^{110}Cd	^{111}In	^{112}Sn	^{113}Sb	^{114}Te	^{115}I	^{116}Xe	^{117}Cs					
63	^{109}Pd	^{110}Ag	^{111}Cd	^{112}In	^{113}Sn	^{114}Sb	^{115}Te	^{116}I	^{117}Xe	^{118}Cs	^{119}Ba	57			
64	^{110}Pd	^{111}Ag	^{112}Cd	^{113}In	^{114}Sn	^{115}Sb	^{116}Te	^{117}I	^{118}Xe	^{119}Cs		La			
65	^{111}Pd	^{112}Ag	^{113}Cd	^{114}In	^{115}Sn	^{116}Sb	^{117}Te	^{118}I	^{119}Xe	^{120}Cs	^{121}Ba				
66	^{112}Pd	^{113}Ag	^{114}Cd	^{115}In	^{116}Sn	^{117}Sb	^{118}Te	^{119}I	^{120}Xe	^{121}Cs	^{122}Ba				
67	^{113}Pd	^{114}Ag	^{115}Cd	^{116}In	^{117}Sn	^{118}Sb	^{119}Te	^{120}I	^{121}Xe	^{122}Cs	^{123}Ba		58		
68	^{114}Pd	^{115}Ag	^{116}Cd	^{117}In	^{118}Sn	^{119}Sb	^{120}Te	^{121}I	^{122}Xe	^{123}Cs	^{124}Ba	^{125}La	Ce		
69	^{115}Pd	^{116}Ag	^{117}Cd	^{118}In	^{119}Sn	^{120}Sb	^{121}Te	^{122}I	^{123}Xe	^{124}Cs	^{125}Ba	^{126}La			
70	^{116}Pd	^{117}Ag	^{118}Cd	^{119}In	^{120}Sn	^{121}Sb	^{122}Te	^{123}I	^{124}Xe	^{125}Cs	^{126}Ba	^{127}La		59	
71	^{117}Pd	^{118}Ag	^{119}Cd	^{120}In	^{121}Sn	^{122}Sb	^{123}Te	^{124}I	^{125}Xe	^{126}Cs	^{127}Ba	^{128}La	^{129}Ce	Pr	60
72	^{118}Pd	^{119}Ag	^{120}Cd	^{121}In	^{122}Sn	^{123}Sb	^{124}Te	^{125}I	^{126}Xe	^{127}Cs	^{128}Ba	^{129}La	^{130}Ce		Nd
73		^{120}Ag	^{121}Cd	^{122}In	^{123}Sn	^{124}Sb	^{125}Te	^{126}I	^{127}Xe	^{128}Cs	^{129}Ba	^{130}La	^{131}Ce	^{132}Pr	
74		^{121}Ag	^{122}Cd	^{123}In	^{124}Sn	^{125}Sb	^{126}Te	^{127}I	^{128}Xe	^{129}Cs	^{130}Ba	^{131}La	^{132}Ce	^{133}Pr	^{134}Nd
75		^{122}Ag		^{124}In	^{125}Sn	^{126}Sb	^{127}Te	^{128}I	^{129}Xe	^{130}Cs	^{131}Ba	^{132}La	^{133}Ce	^{134}Pr	^{135}Nd
76		^{123}Ag	^{124}Cd	^{125}In	^{126}Sn	^{127}Sb	^{128}Te	^{129}I	^{130}Xe	^{131}Cs	^{132}Ba	^{133}La	^{134}Ce	^{135}Pr	^{136}Nd
77	77			^{126}In	^{127}Sn	^{128}Sb	^{129}Te	^{130}I	^{131}Xe	^{132}Cs	^{133}Ba	^{134}La	^{135}Ce	^{136}Pr	^{137}Nd
78		78		^{127}In	^{128}Sn	^{129}Sb	^{130}Te	^{131}I	^{132}Xe	^{133}Cs	^{134}Ba	^{135}La	^{136}Ce	^{137}Pr	^{138}Nd
79		79		^{128}In	^{129}Sn	^{130}Sb	^{131}Te	^{132}I	^{133}Xe	^{134}Cs	^{135}Ba	^{136}La	^{137}Ce	^{138}Pr	^{139}Nd
80		80		^{129}In	^{130}Sn	^{131}Sb	^{132}Te	^{133}I	^{134}Xe	^{135}Cs	^{136}Ba	^{137}La	^{138}Ce	^{139}Pr	^{140}Nd
81		81		^{130}In	^{131}Sn	^{132}Sb	^{133}Te	^{134}I	^{135}Xe	^{136}Cs	^{137}Ba	^{138}La	^{139}Ce	^{140}Pr	^{141}Nd
82		82		^{131}In	^{132}Sn	^{133}Sb	^{134}Te	^{135}I	^{136}Xe	^{137}Cs	^{138}Ba	^{139}La	^{140}Ce	^{141}Pr	^{142}Nd
83		83		^{132}In	^{133}Sn	^{134}Sb	^{135}Te	^{136}I	^{137}Xe	^{138}Cs	^{139}Ba	^{140}La	^{141}Ce	^{142}Pr	^{143}Nd
84			84		^{134}Sn	^{135}Sb	^{136}Te	^{137}I	^{138}Xe	^{139}Cs	^{140}Ba	^{141}La	^{142}Ce	^{143}Pr	^{144}Nd
85				85		^{136}Sb	^{137}Te	^{138}I	^{139}Xe	^{140}Cs	^{141}Ba	^{142}La	^{143}Ce	^{144}Pr	^{145}Nd
86					86		^{138}Te	^{139}I	^{140}Xe	^{141}Cs	^{142}Ba	^{143}La	^{144}Ce	^{145}Pr	^{146}Nd
87						87		^{140}I	^{141}Xe	^{142}Cs	^{143}Ba	^{144}La	^{145}Ce	^{146}Pr	^{147}Nd
88						88		^{141}I	^{142}Xe	^{143}Cs	^{144}Ba	^{145}La	^{146}Ce	^{147}Pr	^{148}Nd
89						89		^{142}I	^{143}Xe	^{144}Cs	^{145}Ba	^{146}La	^{147}Ce	^{148}Pr	^{149}Nd
90							90		^{144}Xe	^{145}Cs	^{146}Ba	^{147}La	^{148}Ce	^{149}Pr	^{150}Nd
91							91		^{145}Xe	^{146}Cs		^{148}La	^{149}Ce	^{150}Pr	^{151}Nd
92											92		^{150}Ce	^{151}Pr	^{152}Nd
93											93		^{151}Ce		
94											94				^{154}Nd

Isotopes for elements 61-75

p →	61 Pm	62 Sm	63 Eu	64 Gd	65 Tb	66 Dy	67 Ho	68 Er	69 Tm	70 Yb	71 Lu	72 Hf	73 Ta	74 W	75 Re
n ↓		^{137}Sm		64											
76	^{137}Pm	^{138}Sm	^{139}Eu	Gd											
77	^{138}Pm	^{139}Sm	^{140}Eu		65										
78	^{139}Pm	^{140}Sm	^{141}Eu	^{142}Gd	Tb	66									
79	^{140}Pm	^{141}Sm	^{142}Eu	^{143}Gd		Dy	67	68	69						
80	^{141}Pm	^{142}Sm	^{143}Eu	^{144}Gd			Ho	Er	Tm	70					
81	^{142}Pm	^{143}Sm	^{144}Eu	^{145}Gd	^{146}Tb	^{147}Dy				Yb	71				
82	^{143}Pm	^{144}Sm	^{145}Eu	^{146}Gd	^{147}Tb	^{148}Dy		^{150}Er	^{151}Tm		Lu	72			
83	^{144}Pm	^{145}Sm	^{146}Eu	^{147}Gd	^{148}Tb	^{149}Dy	^{150}Ho	^{151}Er				Hf			
84	^{145}Pm	^{146}Sm	^{147}Eu	^{148}Gd	^{149}Tb	^{150}Dy	^{151}Ho	^{152}Er	^{153}Tm	^{154}Yb	^{155}Lu				
85	^{146}Pm	^{147}Sm	^{148}Eu	^{149}Gd	^{150}Tb	^{151}Dy	^{152}Ho	^{153}Er	^{154}Tm	^{155}Yb	^{156}Lu	^{157}Hf	73	74	
86	^{147}Pm	^{148}Sm	^{149}Eu	^{150}Gd	^{151}Tb	^{152}Dy	^{153}Ho	^{154}Er	^{155}Tm	^{156}Yb		^{158}Hf	Ta	W	
87	^{148}Pm	^{149}Sm	^{150}Eu	^{151}Gd	^{152}Tb	^{153}Dy	^{154}Ho	^{155}Er	^{156}Tm	^{157}Yb		^{159}Hf			
88	^{149}Pm	^{150}Sm	^{151}Eu	^{152}Gd	^{153}Tb	^{154}Dy	^{155}Ho	^{156}Er	^{157}Tm	^{158}Yb		^{160}Hf		^{162}W	
89	^{150}Pm	^{151}Sm	^{152}Eu	^{153}Gd	^{154}Tb	^{155}Dy	^{156}Ho	^{157}Er	^{158}Tm			^{161}Hf		^{163}W	
90	^{151}Pm	^{152}Sm	^{153}Eu	^{154}Gd	^{155}Tb	^{156}Dy	^{157}Ho	^{158}Er	^{159}Tm	^{160}Yb	^{161}Lu			^{164}W	75
91	^{152}Pm	^{153}Sm	^{154}Eu	^{155}Gd	^{156}Tb	^{157}Dy	^{158}Ho	^{159}Er	^{160}Tm	^{161}Yb	^{162}Lu			^{165}W	Re
92	^{153}Pm	^{154}Sm	^{155}Eu	^{156}Gd	^{157}Tb	^{158}Dy	^{159}Ho	^{160}Er	^{161}Tm	^{162}Yb				^{166}W	
93	^{154}Pm	^{155}Sm	^{156}Eu	^{157}Gd	^{158}Tb	^{159}Dy	^{160}Ho	^{161}Er	^{162}Tm	^{163}Yb	^{164}Lu		^{166}Ta		
94		^{156}Sm	^{157}Eu	^{158}Gd	^{159}Tb	^{160}Dy	^{161}Ho	^{162}Er	^{163}Tm	^{164}Yb	^{165}Lu	^{166}Hf	^{167}Ta		
95		^{157}Sm	^{158}Eu	^{159}Gd	^{160}Tb	^{161}Dy	^{162}Ho	^{163}Er	^{164}Tm	^{165}Yb	^{166}Lu	^{167}Hf	^{168}Ta		^{170}Re
96			^{159}Eu	^{160}Gd	^{161}Tb	^{162}Dy	^{163}Ho	^{164}Er	^{165}Tm	^{166}Yb	^{167}Lu	^{168}Hf	^{169}Ta	^{170}W	
97			^{160}Eu	^{161}Gd	^{162}Tb	^{163}Dy	^{164}Ho	^{165}Er	^{166}Tm	^{167}Yb	^{168}Lu	^{169}Hf	^{170}Ta	^{171}W	^{172}Re
98				^{162}Gd	^{163}Tb	^{164}Dy	^{165}Ho	^{166}Er	^{167}Tm	^{168}Yb	^{169}Lu	^{170}Hf	^{171}Ta	^{172}W	
99					^{164}Tb	^{165}Dy	^{166}Ho	^{167}Er	^{168}Tm	^{169}Yb	^{170}Lu	^{171}Hf	^{172}Ta	^{173}W	^{174}Re
100						^{166}Dy	^{167}Ho	^{168}Er	^{169}Tm	^{170}Yb	^{171}Lu	^{172}Hf	^{173}Ta	^{174}W	^{175}Re
101						^{167}Dy	^{168}Ho	^{169}Er	^{170}Tm	^{171}Yb	^{172}Lu	^{173}Hf	^{174}Ta	^{175}W	^{176}Re
102							^{169}Ho	^{170}Er	^{171}Tm	^{172}Yb	^{173}Lu	^{174}Hf	^{175}Ta	^{176}W	^{177}Re
103							^{170}Ho	^{171}Er	^{172}Tm	^{173}Yb	^{174}Lu	^{175}Hf	^{176}Ta	^{177}W	^{178}Re
104								^{172}Er	^{173}Tm	^{174}Yb	^{175}Lu	^{176}Hf	^{177}Ta	^{178}W	^{179}Re
105								^{173}Er	^{174}Tm	^{175}Yb	^{176}Lu	^{177}Hf	^{178}Ta	^{179}W	^{180}Re
106									^{175}Tm	^{176}Yb	^{177}Lu	^{178}Hf	^{179}Ta	^{180}W	^{181}Re
107									^{176}Tm	^{177}Yb	^{178}Lu	^{179}Hf	^{180}Ta	^{181}W	^{182}Re
108										^{178}Yb	^{179}Lu	^{180}Hf	^{181}Ta	^{182}W	^{183}Re
109											^{180}Lu	^{181}Hf	^{182}Ta	^{183}W	^{184}Re
110												^{182}Hf	^{183}Ta	^{184}W	^{185}Re
111												^{183}Hf	^{184}Ta	^{185}W	^{186}Re
112												^{184}Hf	^{185}Ta	^{186}W	^{187}Re
113												^{185}Hf	^{186}Ta	^{187}W	^{188}Re
114														^{188}W	^{189}Re
115														^{189}W	^{190}Re
116														^{190}W	^{191}Re
117															^{192}Re

Isotopes for elements 76-90

n ↓	76 Os	77 Ir	78 Pt	79 Au	80 Hg	81 Tl	82 Pb	83 Bi	84 Po	85 At	86 Rn	87 Fr	88 Ra	89 Ac	90 Th
93	^{169}Os														
94	^{170}Os	^{171}Ir													
95	^{171}Os	^{172}Ir	^{173}Pt												
96	^{172}Os	^{171}Ir	^{174}Pt	^{175}Au											
97	^{173}Os	^{174}Ir	^{175}Pt	^{176}Au	^{177}Hg										
98	^{174}Os	^{175}Ir	^{176}Pt	^{177}Au	^{178}Hg										
99	^{175}Os	^{176}Ir	^{177}Pt	^{178}Au	^{179}Hg										
100	^{176}Os	^{177}Ir	^{178}Pt	^{179}Au	^{180}Hg										
101	^{177}Os	^{178}Ir	^{179}Pt		^{181}Hg										
102	^{178}Os	^{179}Ir	^{180}Pt	^{181}Au	^{182}Hg										
103	^{179}Os	^{180}Ir	^{181}Pt	^{182}Au	^{183}Hg	^{184}Tl	^{185}Pb								
104	^{180}Os	^{181}Ir	^{182}Pt	^{183}Au	^{184}Hg	^{185}Tl	^{186}Pb								
105	^{181}Os	^{182}Ir	^{183}Pt	^{184}Au	^{185}Hg	^{186}Tl	^{187}Pb								
106	^{182}Os	^{183}Ir	^{184}Pt	^{185}Au	^{186}Hg	^{187}Tl	^{188}Pb	^{189}Bi							
107	^{183}Os	^{184}Ir	^{185}Pt	^{186}Au	^{187}Hg	^{188}Tl	^{189}Pb	^{190}Bi							
108	^{184}Os	^{185}Ir	^{186}Pt	^{187}Au	^{188}Hg	^{189}Tl	^{190}Pb	^{191}Bi							
109	^{185}Os	^{186}Ir	^{187}Pt	^{188}Au	^{189}Hg	^{190}Tl	^{191}Pb	^{192}Bi	^{193}Po	^{194}At					
110	^{186}Os	^{187}Ir	^{188}Pt	^{189}Au	^{190}Hg	^{191}Tl	^{192}Pb	^{193}Bi	^{194}Po	^{195}At					
111	^{187}Os	^{188}Ir	^{189}Pt	^{190}Au	^{191}Hg	^{192}Tl	^{193}Pb	^{194}Bi	^{195}Po	^{196}At					
112	^{188}Os	^{189}Ir	^{190}Pt	^{191}Au	^{192}Hg	^{193}Tl	^{194}Pb	^{195}Bi	^{196}Po	^{197}At					
113	^{189}Os	^{190}Ir	^{191}Pt	^{192}Au	^{193}Hg	^{194}Tl	^{195}Pb	^{196}Bi	^{197}Po	^{198}At					
114	^{190}Os	^{191}Ir	^{192}Pt	^{193}Au	^{194}Hg	^{195}Tl	^{196}Pb	^{197}Bi	^{198}Po	^{199}At	^{200}Rn				
115	^{191}Os	^{192}Ir	^{193}Pt	^{194}Au	^{195}Hg	^{196}Tl	^{197}Pb	^{198}Bi	^{199}Po	^{200}At	^{201}Rn				
116	^{192}Os	^{193}Ir	^{194}Pt	^{195}Au	^{196}Hg	^{197}Tl	^{198}Pb	^{199}Bi	^{200}Po	^{201}At	^{202}Rn	^{203}Fr			
117	^{193}Os	^{194}Ir	^{195}Pt	^{196}Au	^{197}Hg	^{198}Tl	^{199}Pb	^{200}Bi	^{201}Po	^{202}At	^{203}Rn	^{204}Fr			
118	^{194}Os	^{195}Ir	^{196}Pt	^{197}Au	^{198}Hg	^{199}Tl	^{200}Pb	^{201}Bi	^{202}Po	^{203}At	^{204}Rn	^{205}Fr	^{206}Ra		
119	^{195}Os	^{196}Ir	^{197}Pt	^{198}Au	^{199}Hg	^{200}Tl	^{201}Pb	^{202}Bi	^{203}Po	^{204}At	^{205}Rn	^{206}Fr	^{207}Ra		
120	^{196}Os	^{197}Ir	^{198}Pt	^{199}Au	^{200}Hg	^{201}Tl	^{202}Pb	^{203}Bi	^{204}Po	^{205}At	^{206}Rn	^{207}Fr	^{208}Ra	^{209}Ac	
121		^{198}Ir	^{199}Pt	^{200}Au	^{201}Hg	^{202}Tl	^{203}Pb	^{204}Bi	^{205}Po	^{206}At	^{207}Rn	^{208}Fr	^{209}Ra	^{210}Ac	
122			^{200}Pt	^{201}Au	^{202}Hg	^{203}Tl	^{204}Pb	^{205}Bi	^{206}Po	^{207}At	^{208}Rn	^{209}Fr	^{210}Ra	^{211}Ac	
123			^{201}Pt	^{202}Au	^{203}Hg	^{204}Tl	^{205}Pb	^{206}Bi	^{207}Po	^{208}At	^{209}Rn	^{210}Fr	^{211}Ra	^{212}Ac	^{213}Th
124				^{203}Au	^{204}Hg	^{205}Tl	^{206}Pb	^{207}Bi	^{208}Po	^{209}At	^{210}Rn	^{211}Fr	^{212}Ra	^{213}Ac	^{214}Th
125				^{204}Au	^{205}Hg	^{206}Tl	^{207}Pb	^{208}Bi	^{209}Po	^{210}At	^{211}Rn	^{212}Fr	^{213}Ra	^{214}Ac	^{215}Th
126					^{206}Hg	^{207}Tl	^{208}Pb	^{209}Bi	^{210}Po	^{211}At	^{212}Rn	^{213}Fr	^{214}Ra	^{215}Ac	^{216}Th
127						^{208}Tl	^{209}Pb	^{210}Bi	^{211}Po	^{212}At	^{213}Rn	^{214}Fr	^{215}Ra	^{216}Ac	^{217}Th
128						^{209}Tl	^{210}Pb	^{211}Bi	^{212}Po	^{213}At	^{214}Rn	^{215}Fr	^{216}Ra	^{217}Ac	^{218}Th
129						^{210}Tl	^{211}Pb	^{212}Bi	^{213}Po	^{214}At	^{215}Rn	^{216}Fr	^{217}Ra	^{218}Ac	^{219}Th
130							^{212}Pb	^{213}Bi	^{214}Po	^{215}At	^{216}Rn	^{217}Fr	^{218}Ra	^{219}Ac	^{220}Th
131							^{213}Pb	^{214}Bi	^{215}Po	^{216}At	^{217}Rn	^{218}Fr	^{219}Ra	^{220}Ac	^{221}Th
132							^{214}Pb	^{215}Bi	^{216}Po	^{217}At	^{218}Rn	^{219}Fr	^{220}Ra	^{221}Ac	^{222}Th
133									^{217}Po	^{218}At	^{219}Rn	^{220}Fr	^{221}Ra	^{222}Ac	^{223}Th
134									^{218}Po	^{219}At	^{220}Rn	^{221}Fr	^{222}Ra	^{223}Ac	^{224}Th
135											^{221}Rn	^{222}Fr	^{223}Ra	^{224}Ac	^{225}Th
136											^{222}Rn	^{223}Fr	^{224}Ra	^{225}Ac	^{226}Th
137											^{223}Rn	^{224}Fr	^{225}Ra	^{226}Ac	^{227}Th
138											^{224}Rn	^{225}Fr	^{226}Ra	^{227}Ac	^{228}Th
139											^{225}Rn	^{226}Fr	^{227}Ra	^{228}Ac	^{229}Th
140											^{226}Rn	^{227}Fr	^{228}Ra	^{229}Ac	^{230}Th
141												^{228}Fr	^{229}Ra	^{230}Ac	^{231}Th
142												^{229}Fr	^{230}Ra	^{231}Ac	^{232}Th
143														^{232}Ac	^{233}Th
144															^{234}Th
145															^{235}Th
146															^{236}Th

Isotopes for elements 91-105

p → 91 Pa

n ↓

n	91 Pa	92 U	93 Np	94 Pu	95 Am	96 Cm	97 Bk	98 Cf	99 Es	100 Fm	101 Md	102 No	103 Lr	104 Rf	105 Db
125	^{216}Pa														
126															
127															
128															
129															
130															
131	^{222}Pa	92													
132	^{223}Pa	U	93												
133	^{224}Pa		Np	94											
134	^{225}Pa	^{226}U		Pu	95										
135	^{226}Pa	^{227}U	^{228}Np		Am										
136	^{227}Pa	^{228}U	^{229}Np			96									
137	^{228}Pa	^{229}U	^{230}Np		^{232}Am	Cm									
138	^{229}Pa	^{230}U	^{231}Np	^{232}Pu											
139	^{230}Pa	^{231}U	^{232}Np	^{233}Pu	^{234}Am		97	98	99	100					
140	^{231}Pa	^{232}U	^{233}Np	^{234}Pu	^{235}Am	^{236}Cm	Bk	Cf	Es	Fm					
141	^{232}Pa	^{233}U	^{234}Np	^{235}Pu	^{236}Am	^{237}Cm									
142	^{233}Pa	^{234}U	^{235}Np	^{236}Pu	^{237}Am	^{238}Cm	^{239}Bk	^{240}Cf	^{241}Es	^{242}Fm					
143	^{234}Pa	^{235}U	^{236}Np	^{237}Pu	^{238}Am	^{239}Cm	^{240}Bk	^{241}Cf	^{242}Es	^{243}Fm					
144	^{235}Pa	^{236}U	^{237}Np	^{238}Pu	^{239}Am	^{240}Cm	^{241}Bk	^{242}Cf	^{243}Es	^{244}Fm	101				
145	^{236}Pa	^{237}U	^{238}Np	^{239}Pu	^{240}Am	^{241}Cm	^{242}Bk	^{243}Cf	^{244}Es	^{245}Fm	Md	102			
146	^{237}Pa	^{238}U	^{239}Np	^{240}Pu	^{241}Am	^{242}Cm	^{243}Bk	^{244}Cf	^{245}Es	^{246}Fm		No	103	104	
147	^{238}Pa	^{239}U	^{240}Np	^{241}Pu	^{242}Am	^{243}Cm	^{244}Bk	^{245}Cf	^{246}Es	^{247}Fm	^{248}Md		Lr	Rf	105
148		^{240}U	^{241}Np	^{242}Pu	^{243}Am	^{244}Cm	^{245}Bk	^{246}Cf	^{247}Es	^{248}Fm	^{249}Md	^{250}No			Db
149	149		^{243}Pu	^{244}Am	^{245}Cm	^{246}Bk	^{247}Cf	^{248}Es	^{249}Fm	^{250}Md	^{251}No	^{252}Lr	^{253}Rf		
150		150	^{244}Pu	^{245}Am	^{246}Cm	^{247}Bk	^{248}Cf	^{249}Es	^{250}Fm	^{251}Md	^{252}No	^{253}Lr	^{254}Rf	^{255}Db	
151		151	^{245}Pu	^{246}Am	^{247}Cm	^{248}Bk	^{249}Cf	^{250}Es	^{251}Fm	^{252}Md	^{253}No	^{254}Lr	^{255}Rf	^{256}Db	
152		152	^{246}Pu	^{247}Am	^{248}Cm	^{249}Bk	^{250}Cf	^{251}Es	^{252}Fm	^{253}Md	^{254}No	^{255}Lr	^{256}Rf	^{257}Db	
153			153	^{249}Cm	^{250}Bk	^{251}Cf	^{252}Es	^{253}Fm	^{254}Md	^{255}No	^{256}Lr	^{257}Rf	^{258}Db		
154				154	^{250}Cm	^{251}Bk	^{252}Cf	^{253}Es	^{254}Fm	^{255}Md	^{256}No	^{257}Lr	^{258}Rf	^{259}Db	
155				155	^{251}Cm	^{252}Bk	^{253}Cf	^{254}Es	^{255}Fm	^{256}Md	^{257}No	^{258}Lr	^{259}Rf	^{260}Db	
156					156	^{253}Bk	^{254}Cf	^{255}Es	^{256}Fm	^{257}Md	^{258}No	^{259}Lr	^{260}Rf	^{261}Db	
157					157	^{254}Bk	^{255}Cf	^{256}Es	^{257}Fm	^{258}Md	^{259}No	^{260}Lr	^{261}Rf	^{262}Db	
158						158	^{256}Cf	^{257}Es	^{258}Fm	^{259}Md	^{260}No	^{261}Lr	^{262}Rf	^{263}Db	
159							159	^{259}Fm	^{260}Md	^{261}No	^{262}Lr	^{263}Rf	^{264}Db		
160								160	^{262}No	^{263}Lr	^{264}Rf	^{265}Db			

Isotopes for elements 106-111

n ↓	Sg (106)	Bh (107)	Hs (108)	Mt (109)	Ds (110)	Rg (111)
152	^{258}Sg		108			
153	^{259}Sg	^{260}Bh	Hs	109		
154	^{260}Sg	^{261}Bh		Mt	110	
155	^{261}Sg	^{262}Bh	^{263}Hs		Ds	111
156	^{262}Sg	^{263}Bh	^{264}Hs	^{265}Mt		Rg
157	^{263}Sg	^{264}Bh	^{265}Hs	^{266}Mt	^{267}Ds	
158	^{264}Sg	^{265}Bh	^{266}Hs	^{267}Mt	^{268}Ds	
159	^{265}Sg	^{266}Bh	^{267}Hs	^{268}Mt	^{269}Ds	
160	^{266}Sg	^{267}Bh	^{268}Hs	^{269}Mt	^{270}Ds	
161			^{269}Hs	^{270}Mt	^{271}Ds	^{272}Rg
162				^{271}Mt	^{272}Ds	

http://www.answers.com/topic/isotope-table
http://hyperphysics.phy-astr.gsu.edu/hbase/nuclear/nucstructcon.html

****
****

THE FIRST COVER

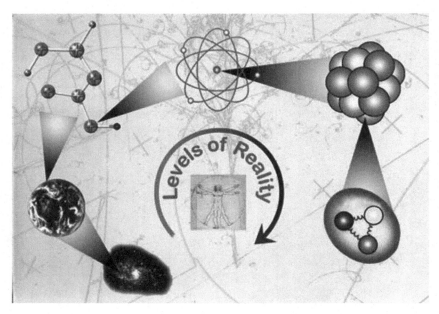

The hierarchy of decreasing dimensions, from galaxies (lower left) to quarks (lower right) through the earth, molecules, atoms and nuclei. Nuclear Physics is relevant to the description of all these "objects".

http://www.finuphy.org/

Mouaiyad Alabed

1987 - 1992 - BGY Belarusian Stat University.
Ph.D in Technical Science
1985 - 1987 - BGY Belarusian Stat University.
(Master of science) in physics and mathematics

1978 - 1982   Baghdad University   Iraq
Bsc. In Physics
1992-2001   Senior Lectors in Science Faculty. Division of Physics in the Several
Universities.
¤   Chef of the division of Physics in Science Faculty.
     Altahaddi University in Libya.

¤   Chef of the division of high studies in Science Faculty.
     Altahaddi University in Libya.

¤   Director for various students projects for ultimate years students.

¤   Reviewer various research for that attain Mc in Science Faculty. Nasir
University
¤   Published two books:
"Physical and Geological phenomena in Holey Quraan" in Arabic.
 "English Arabic Dictionary in Nuclear Physics".
¤ Three books is completed for publishing:
"Questions and Answers in Nuclear Energy"

 " Physical Dictionary" English Arabic dictionary.
"Problems and Solution in Nuclear Physics" You can see it!
¤ Editor for   ALFIZIAII ` Fysisten` Altahaddi University Libya

¤ Lecturer of various seminar about radiation contamination in different places in the
world.

¤ Writing of the different articles of the radiation contamination in different
magazines.

¤ Member with the staff of Department of physics – Lund University as a trainee in
Lunds University with AMS (accelerator mass spectrometry) group.

¤ Practice in Lunds University. Camp of Helsingborg –Technique College.
 I have two research on midst precinct unpublished.

¤ Several papers in my field.